Origin and Evolution of the Atmosphere and Ocean

Guido Visconti

Origin and Evolution of the Atmosphere and Ocean

Springer

Guido Visconti
Department of Physical and Chemical Sciences
Università dell'Aquila
L'Aquila, Italy

ISBN 978-3-032-02652-1 ISBN 978-3-032-02653-8 (eBook)
https://doi.org/10.1007/978-3-032-02653-8

This Springer imprint is published by the registered company Springer Nature Switzerland AG
The registered company address is: Gewerbestrasse 11, 6330 Cham, Switzerland

This book is dedicated to the memory and love of my parents Angelina and Ugo and to my only surviving sister Pinuccia.

Preface

My academic career in terms of publications began possibly 50 years ago with a paper published in the *Journal of the Atmospheric Sciences* on the exospheric temperature of the primitive Earth. Then two years later James Walker published his book *Evolution of the Atmosphere* to which the present book is inspired. The subject of the book involves a very large amount of material and spans an impressive number of sectors from radiative theory to geochemestry, from dating methods to climatic theories. Another characteristic of this subject is that the experimental evidence is so sparse and very much indirect so that it is fertile ground for the most fascinating theories with major examples dinosaurs extinction or snowball Earth: that means pure academic. But that also means that could be entertaining and some time very technical. I decided as always to relegate the strictly technical details and annoying derivaation to the appendix where also some MATLAB program is presented. Considering the epoch in which appeared, half of the Walker book was dedicated to the description of the present atmosphere that we relegate just to Chap. 8. Otherwise we start with a description of the solar system because from its basic data we can extrapolate a few ideas on its origin. Once the formation of the solar system is outlined we describe the formation of the Earth and its thermal and chemical evolution. Here we have to deal with some celestial mechanics to describe formation and dynamics of the planetesimals, the building blocks of planets. The formation and growth of the Earth is described both for the "solid part" and later for its chemical and minearological aspects. Here we begin to deal with the birth of the primordial atmosphere where the examination of the noble gas inventory may give important clues on the origin of the present atmosphere whose composition is largely determined by the escape mechanisms. These are examined in great details in Chap. 3. The next chapter is dedicated almost entirely to the primordial reducing atmosphere. Chapter 5 is dedicated to the origin of the oceans. I found here very sparse material on the real origin of the ocean while there are so many reference to the present status of the oceans. The main distimnction of the Earth's atmosphere is the presence of oxygen and its very complex interaction with life. The raise of oxygen is the main topic of Chap. 6 which has to be treated in the framework of the biogeochenical cycles. The stability of the oxygen is mostly mistreated in the holy texts and wrongly attributed simply to the presence of photosynthesis neglecting the complex interaction with the rock cycle. In the classical Walker text curiosly enough there is no mention of the climate evolution on Earth although both atmosphere and oceans are components of the climate system. This topic is developed in Chap. 7 starting with a review of the methods to measure ancient temperature and taking enough space to deal with the faint sun paradox, the snowball Earth, the dinosaur extinction up to the ice ages. Then in Chap. 8 the present status of the stmosphere and ocean is reviewed discussing the problems of the human influence on their composition. In conclusion Chap. 9 is dedicated to the future of the atmosphere and ocean. Here we mention two important recent developments that is the "tippig point" and the concept of the "great filter" originated from astrobiology in the discussions of the Fermi paradox. Both these concepts could influence suddently the evolution of the atmosphere and oceans whose final destiny is connected in any case to the evolution of our star: the Sun. As you can imagine we are crossing the border of astrobiology which a very fascinating realm. I have tried some time to lightnen the academic tone of the book considering that a few of the theories and constructs are themselves an expression of the academic environment: you change a constant a made another publication or a publications is made without comparing the results with the experiments or the data. Some time it happens that a few results are plainly wrong but that is in the inner weels of the system. In the same line of reasoning I have mentioned only a few times the anthropocene because I think it is a tipical academic invention.

L'Aquila, Italy Guido Visconti

Acknowledgements A large number of references on the subject have appeared in the literature many years ago sometime in obscure journals, and Antonella Di Nisio was fundamental to recover them and so deserve my acknowledgements considering also her new responsibilities in the library system of my University. Francesco Mulargia of the University of Bologna read the manuscript and was always encouraging as it was Rolando Garcia of the National Center for Atmospheric Research that read and commented a few chapters. The same job was carried out enthusiastically by my collegue Zurab Berejani. My old friend and collegue Marco Verdecchia also commented in details many chapters. Contacts with other collegues have to be acknowledged like Jim Kasting, Michael Ghil and Daniel Jacob. I have dedicated the book tomy early literary passions which include Bob Dylan, Robert Schecley and John Updike with Karl Barth.

Competing Interests The author has no competing interests to declare that are relevant to the content of this manuscript.

Contents

Nomenclature

g	Acceleration of gravity
$I_{1/4}$	Modified bessel function
z	Altitude
ω, Ω	Angular velocity, vorticity, wave frequency
H	Scale height
N	Buoyancy frequency
μ	Coefficient of viscosity
ρ	Density
D	Diffusion coefficient
G	Gravitational constant
E	Energy, electric field
R	Gas constant
G	Gravitational constant
c_s	Sound velocity
ν	Kinematic viscosity
μ	Dynamic viscosity
$J_\odot$	Angular momentum of the sun
$M_\odot$	Mass of the sun
Ω_K	Keplerian angular velocity
ϕ	Potential, velocity potential
U	Potential energy
Σ	Surface density
p	Pressure
$R_\odot$	Radius of the sun
Ra	Rayleigh number
Nu	Nusselt number
c_p	Specific heat at constant pressure
c_v	Specific heat at constant volume
k	Boltzmann constant, absorption coefficient
m_p	Mass of the fragment
α	Albedo

S	Salinity
F_g	Focusing factor
Ψ_N	Angular momentum
Q	Toomre parameter
F_{grav}	Vertical gravity
F_D	Drag force
t_ν	Viscous time
v_{set}	Settling velocity
r_H	Hill radius
τ	Stress
τ_ν	Viscous stress
t_ν	Viscous time
T	Total torque
T	Temperature
v_ϕ	Tangential velocity, azimutal velocity
v_r	Radial velocity
V_i	Impact velocity
T_p	Planetesimal temperature
λ	Wavelength
M_w	Mass of water vapor
T_{sol}	Solidus temperature
E_R	Radioactive heat generation
F_0	Incoming solar flux
κ	Diffusion coefficent
Φ_T	Thermal energy flux
v_e	Escape velocity
U	Most probable thermal velocity
Phi_j	Jeans escape flux
Φ_l	Limiting flux
$M_\oplus$	Mass of the Earth
$R_\oplus$	Radius of the Earth
P_{CO_2}	Carbon dioxide pressure
K_{oxy}	Oxygenation parameter
Φ_{out}	Outgassing flux
Φ_{OW}	Oxidative flux
Φ_{burial}	Burial rate
F_{org}	Organic carbon burial
W	Weathering rate
V	Emission rate from volcanoes
Q_e	Extinction effivciency
ω	Single scattering albedo
F^-	Downwad diffuse flux
F^+	Upnwad diffuse flux

Chapter 1
The Solar System

1.1 Solar System Planets

To understand the origin and evolution of the atmosphere and oceans on our planet (the Earth) we need to give some data on our solar system. As we know it is made up by the Sun, eight planets, the asteroid belt and some minor object. **Table** 1.1 give a summary of some of the data relevant to the planets and the Sun.

A few observations can be made from this table. The mass of the Sun represents almost the total mass and the orbits of the planets are circular except for Mercury whose orbit is probably perturbed by the vicinity of the Sun. The same is true for the inclination, all the orbits lies essentially on the ecliptic plane except again for Mercury. Notice also the sudden drop in average density after Mars a data which reveals a quite different composition between the inner planets (Mercury, Venus, Earth and Mars) called *terrestrial* and the *Jovian* planets (Jupiter, Saturn, Uranus and Neptune). In the table we do not report Pluto, the asteroid belt and some trans neptunian object like Ceres. We can calculate another data which is the angular momentum. We consider the Sun as a sphere with radius $R_\odot$ and mass $M_\odot$ then the angular momentum if the Sun rotates with angular velocity ω is given by

$$J_\odot = kM_\odot R_\odot^2 \tag{1.1}$$

And using the data in Table 1.1 with $\Omega \approx 2.59 \times 10^{-6}$ sec^{-1} (corresponding to an average period of 28 days) we have $J_\odot \approx 2.68 \times 10^{41}$ kg m^2. For a solid sphere $k = 0.4$ but the Sun is not solid so we used the accepted value of 0.1. On the other hand if we consider the largest planet (Jupiter) its orbital angular momentum is given by

$$J_J = M_J a^2 \Omega_J = M_J \sqrt{(GM_\odot/a)} \approx 2 \times 10^{43} \text{ kg m}^2 \tag{1.2}$$

where G is the gravitational constant and we have used the Keplero law.

G. Visconti, *Origin and Evolution of the Atmosphere and Ocean*,
https://doi.org/10.1007/978-3-032-02653-8_1

Table 1.1 Data on the planets including the Sun. The data refer to the distance from the Sun a in astronomical units (A.U.), the eccentricity of the orbit, e, the inclination on the ecliptic plane i, the mass, M, the radius R and the average density ρ

	a(AU)	e	i (deg)	M (kg)	$R(10^3$ km)	ρ (g/cm^3)
Sun				1.99×10^{30}	695.7	1.40
Mercury	0.387	0.205	7.00	3.30×10^{23}	2.44	5.4
Venus	0.723	0,007	3.39	4.87×10^{24}	6.05	5.2
Earth	1.00	0.017	0.00	5.97×10^{24}	6.34	5.5
Mars	1.52	0.093	1.85	6.42×10^{23}	3,39	3.9
Jupiter	5.203	0.048	1.30	1.90×10^{27}	71.49	1.3
Saturn	9–53	0.053	2.49	5.68×10^{26}	60.27	0.7
Uranus	19.19	0.047	0.77	8.68×10^{25}	25.56	1.3
Neptune	30.07	0.009	1.77	1.02×10^{26}	24.76	1.6

The data that constrains all the theories on the origin of the solar system and the origin of the atmospheres are very much evident and may be summarized as follows:

1. The orbit of the planets and most of their satellites are on the same plane. The planets also rotate all in the same direction (anti-clockwise if observed from the north pole of the Sun).
2. Almost all the mass of the solar system (99.9%) is concentrated in the Sun while the angular momentum is concentrated (99.4%) in the planets.
3. All the gravitational energy is concentrated in the Sun. Only one part in a million is found on the planets. This means that a body outside the solar system feels only the attraction of the Sun.
4. There is a net difference in the chemical composition between the planets inside and outside the asteroid belt that separates the orbits of Mars and Jupiter. The internal planets (Mercury, Venus, Earth and Mars) are rocky with densities higher than the density of water and all have a solid surface. These planets are also called *terrestrial*. The outermost planets (Jupiter, Saturn, Uranus and Neptune) have a gaseous aspect and their density and other measurements indicate that their composition is essentially helium and hydrogen. These planets are called *Jovian*.
5. The internal structure of the planets, terrestrial or Jovian, present a differentiation. Even if details change, the terrestrial planets, for example, have their innermost parts made up of heavy minerals while light minerals make up the outermost part.

These data indicate that the solar system probably originated from a single body in rotation whose radial dimension was much larger than its axial (vertical) dimension. In this primordial body (that had the appearance of a disk) there was a progressive transfer of mass toward the center and a migration of angular momentum toward the edge. With the transfer of the mass toward the center, this region was decelerating while the external region was accelerating because of the conservation of angular momentum.

The different chemical compositions of the planets indicate a different thermal regime between the outer part and the central part of the primordial disk. The region near the axis reached higher temperatures than the external part, and in it the first minerals with high fusion point condensed. These were magnesium and iron silicates. In the region toward the rim of the disk, which was much colder, ices (which are also lighter) condensed.

Finally the well-differentiated internal structure of the planets suggests a gravitational separation, with the heaviest mineral (iron and nickel) migrating toward the center of the planets. In very simple terms the accepted theory on the origin of the solar system imitates what we learned in school as Kant and Laplace theories that were formulated between 1755 (Kant) and 1796 (Laplace). In this scheme a huge cloud of dust and gas contracts and rotates forming a disk that has roughly the same dimension of the present solar system. The viscosity in the disk in part dissipates the angular momentum, raising its temperature until a thermal gradient is established along the radial dimension, with higher temperatures at the center. When the disk began to cool, the first "drops" from condensation formed with chemical composition that reflected the temperature of the region. Silicates form near the center of the disk and ices near the rim. The turbulence in the disk was quite high and at some point the density fluctuations gave rise to aggregations of larger dimension and the first grains formed. Collisions among the grains formed the first embryos of the planets (planetesimals). The embryos grew in a chaotic way sweeping everything in their path. The planets were born in this way and those of larger dimensions attracted a major fraction of the gas from the disk. In the planetologist mind, three elements formed the planets: rock, ice and gas. In this simple scheme it is obvious to think that the composition of the terrestrial planets reflects in some way the environment in which they have formed. A proof of that could be found in the comparison of the abundance of the elements on the Sun with the abundance found in a very special class of meteorites called carbonaceous chondrites. However this data refers to elements with very low volatility, that is elements not found easily as a gas under normal conditions of temperature and pressure (on Earth). We need to think that, because the chondrites were processed through the same thermal cycles as the material that makes up the planets, in the latter we could expect some depletion of volatiles. This topic is a very good one to investigate the Early history of the atmosphere so that we will dedicate a large space to it. However the core of the problem is the formation and development of the primitive nebula. In very simple terms the accepted theory on the origin of the solar system imitates what we learned in school as Kant and Laplace theories that were formulated between 1755 (Kant) and 1796 (Laplace). In this scheme a huge cloud of dust and gas contracts and rotates forming a disk that has roughly the same dimension of the present solar system. The viscosity in the disk in part dissipates the angular momentum, raising its temperature until a thermal gradient is established along the radial dimension, with higher temperatures at the center. When the disk began to cool, the first "drops" from condensation formed with chemical composition that reflected the temperature of the region. Silicates form near the center of the disk and ices near the rim. The turbulence in the disk was quite high and at some point the density fluctuations gave rise to aggregations of larger dimension and the first grains formed. Collisions among the grains formed the first embryos of the planets (planetesimals). The embryos grew in a chaotic way sweeping everything in their path. The planets were born in this way and those of larger dimensions attracted a major fraction of the gas from the disk. In the planetologist mind, three elements formed the planets: rock, ice and gas. In this simple scheme it is obvious to think that the composition of the terrestrial planets reflects in some way the environment in which they have formed. A proof of that could be found in the comparison of the abundance of the elements on the Sun with the abundance found in a very special class of meteorites called carbonaceous chondrites. However this data refers to elements with very low volatility, that is elements not found easily as a gas under normal conditions of temperature and pressure (on Earth). We need to think that, because the chondrites were processed through the same thermal cycles as the material that makes up the planets, in the latter we could expect some depletion of volatiles. This topic is a very good one to investigate the Early history of the atmosphere so that we will dedicate a large space to it. However the core of the problem is the formation and development of the primitive nebula. In very simple terms the accepted theory on the origin of the solar system imitates what we learned in school as Kant and Laplace theories that were formulated between 1755 (Kant) and 1796 (Laplace). In this scheme a huge cloud of dust and gas contracts and rotates forming a disk that has roughly the same dimension of the present solar system. The viscosity in the disk in part dissipates the angular momentum, raising its temperature until a thermal gradient is established along the radial dimension, with higher temperatures at the center. When the disk began to cool, the first "drops" from condensation formed with chemical composition that reflected the temperature of the region. Silicates form near the center of the disk and ices near the rim. The turbulence in the disk was quite high and at some point the density fluctuations gave rise to aggregations of larger dimension and the first grains formed. Collisions among the grains formed the first embryos of the planets (planetesimals). The embryos grew in a chaotic way sweeping everything in their path. The planets were born in this way and those of larger dimensions attracted a major fraction of the gas from the disk. In the planetologist mind, three elements formed the planets: rock, ice and gas. In this simple scheme it is obvious to think that the composition of the terrestrial planets reflects in some way the environment in which they have formed. A proof of that could be found in the comparison of the abundance of the elements on the Sun with the abundance found in a very special class of meteorites called carbonaceous chondrites. However this data refers to elements with very low volatility, that is elements not found easily as a gas under normal conditions of temperature and pressure (on Earth). We need to think that, because the chondrites were processed through the same thermal cycles as the material that makes up the planets, in the latter we could expect some depletion of volatiles. This topic is a very good one to investigate the Early history of the atmosphere so that we will dedicate a large space to it. However the core of the problem is the formation and development of the primitive nebula.

1.2 Model for the Primitive Nebula

The problem of the accretion disks is an old one and has been explored initially in connection with the primitive nebula mainly by Donald Lynden-Bell and James Pringle about fifty years ago. Then it became popular in connection with the problem of black-hole after the seminal work of Nikolai Shakura and Rashid Sunyaev. In what follow we will refer mainly to a review paper by Pringle published in 1981. Before that we have to mention what is called the *the minimum mass nebula* as it was invented by Stuart Weidenschilling in a famous 1977 paper. He made a simple calculation.

1. For each planet he considered the mass of heavy elements and added enough hydrogen and helium to reach the solar composition. The mass of Jupiter was less affected by this procedure but not the Earth.
2. He distributed the masses obtained in an annulus around the orbit of the planet to determine the surface density Σ in $g\ cm^{-2}$

Then it results that the density follows the rule $\Sigma \propto r^{-1.5}$. A popular proposal was formulated by Chushiro Hayashi that used the expression

$$\Sigma = 1700 r^{-1.5}\ \mathrm{g\ cm^{-2}} \tag{1.3}$$

where r is expressed in astronomical units (A.U.). Integrating (1.3) up to about 30 A.U. gives a mass around $0.01 M_\odot$ which roughly corresponds to the total mass of the planets.

In order to obtain a first estimation of density distribution within the nebula, let consider a cylindrical coordinate system r, ϕ, z and a disc made up of dust and gas with radial dimension much larger than the vertical dimension. We also assume that the disk rotates with an angular velocity which is variable with r and has a Keplerian value Ω_K given by

$$\Omega_K(r) = \left(\frac{GM}{r^3}\right)^{1/2} \tag{1.4}$$

However the differential rotation of the fluid can be characterized by a shearing A defined as

$$A = r\frac{\partial \Omega}{\partial r} \tag{1.5}$$

We also assume that beside the tangential velocity $v_\phi = r\Omega_k$ we also have a radial drift velocity v_r which is negligible near the center of the disc so that the material can accrete. We further define a surface density Σ

$$\Sigma = \int \rho dz \tag{1.6}$$

Where ρ is the volume density and z the vertical coordinate, Then we consider a portion of the disc between radius r and $r + \Delta r$. We have for the conservation of mass $2\pi r \Delta r \Sigma$ (see Appendix for the equation of continuity)

Fig. 1.1 The viscous torque on a ring

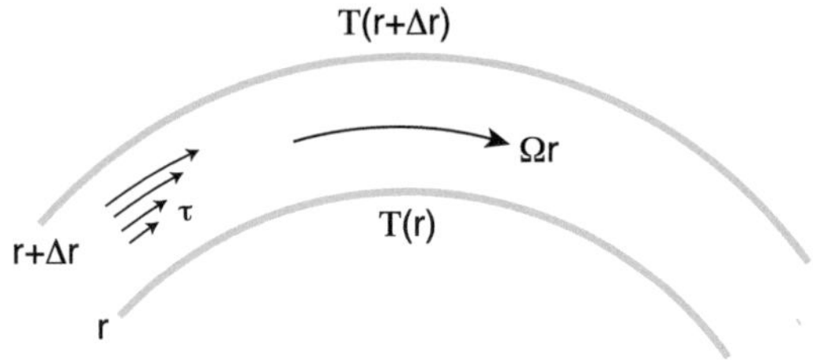

$$r\frac{\partial \Sigma}{\partial t} + \frac{\partial (r\Sigma v_r)}{\partial r} = 0 \tag{1.7}$$

We now consider the angular momentum ΔL of a tiny annular of radius r and width Δr given by

$$\Delta L = \Delta m \Omega r^2 = 2\pi r^2 \Omega \Sigma r \Delta r \tag{1.8}$$

To write down the conservation of angular momentum we need to consider two adjacent annuli with radius r and $r + \Delta r$ and the exchange of angular momentum through the action of viscous torque. If we consider **Fig.** 1.1 we have for the viscous stress

$$\tau_v = \mu \frac{dv}{dr} \approx \mu r \frac{d\Omega}{dr} \tag{1.9}$$

with ν kinematic viscosity. This stress will generate a torque per unit of area $\tau_v r$ and considering the dynamic viscosity $\mu = \nu\Sigma$ we have for the total torque

$$\mathrm{T} = 2\pi r \tau_v r = 2\pi r \nu \Sigma r^2 \frac{d\Omega}{dr} = 2\pi r \nu \Sigma A r \tag{1.10}$$

For the conservation of the angular momentum we make a balance between two adjacent annulus at r and $r + \Delta r$

$$\begin{aligned} 2\pi \frac{\partial}{\partial t}(r^2 \Omega \Sigma r \Delta r) &= 2\pi v_r(r,t) r^2 \Omega(r) \Sigma(r.t) r \Delta r \\ -2\pi v_r(r+\Delta r, t)(r+\Delta r)^2 \Omega(r+\Delta r)\Sigma(r+\Delta r, t) &+ \frac{\partial G}{\partial r}\Delta r \\ \approx -2\pi \Delta r \frac{\partial}{\partial r}(v_r r \Sigma r^2 \Omega) &+ \frac{\partial G}{\partial r}\Delta r \end{aligned} \tag{1.11}$$

Which gives the equation with $\Delta r \to 0$

$$r\frac{\partial}{\partial t}(r^2 \Omega \Sigma) + \frac{\partial}{\partial r}(r \Sigma v_r r^2 \Omega) = \frac{1}{2\pi}\frac{\partial G}{\partial r} \tag{1.12}$$

Now the first term on the left can be split as

$$r\frac{\partial}{\partial t}(r^2\Omega\Sigma) = rr^2\Omega\frac{\partial\Sigma}{\partial t} + rr^2\Sigma\frac{\partial\Omega}{\partial t} = -r^2\frac{\partial(r\Sigma v_r)}{\partial r}$$

So that substituting in (1.12) we have

$$r\Sigma v_r\frac{\partial(\Omega r^2)}{\partial r} = r\Sigma v_r(\Omega r^2)' = \frac{1}{2\pi}\frac{\partial G}{\partial r} \tag{1.13}$$

where ()' indicate derivative with respect to r. We can now use this equation together with (1.7) to eliminate v_r to get

$$r\frac{\partial\Sigma}{\partial t} = -\frac{\partial(r\Sigma v_r)}{\partial r} = -\frac{\partial}{\partial r}\left[\frac{2\pi}{(\Omega r^2)'}\frac{\partial G}{\partial r}\right] \tag{1.14}$$

Now we specialize that the motion be Keplerian so that Ω is given by (1.4) so that all terms can be made explicit and we have

$$\frac{\partial\Sigma}{\partial t} = \frac{3}{r}\frac{\partial}{\partial r}\left[r^{1/2}\frac{\partial(\nu\Sigma r^{1/2})}{\partial r}\right] \tag{1.15}$$

In the same approximation

$$\frac{\partial\Omega}{\partial r} = -\frac{3}{2}\sqrt{GM}r^{-5/2}, \quad G = -3\pi\nu\Sigma\sqrt{GM}r^{1/2}, \quad \frac{\partial G}{\partial r} = -\frac{3}{2}\pi\nu\Sigma\sqrt{GM}r^{-1/2} \tag{1.16}$$

Then from (1.13) we have

$$v_r = \frac{1}{2\pi r\Sigma(\Omega r^2)'}\frac{\partial G}{\partial r} \tag{1.17}$$

where

$$\frac{\partial}{\partial r}(\Omega r^2) = \frac{1}{2}\sqrt{GM}r^{-1/2} \tag{1.18}$$

Finally substituting in (1.17) we have

$$v_r = \frac{3}{\Sigma r^{1/2}}\frac{\partial}{\partial r}(\nu\Sigma r^{1/2}) \tag{1.19}$$

This model does not want to give information about the physical processes which determine the viscosity so we choose $\nu = constant$ and we show in the appendix that a solution in this case is given by

$$\Sigma(x,t) = \frac{m}{\pi r_0\tau x^{1/4}}\exp\left\{\left(\frac{1+x^2}{\tau}\right)\right\}I_{1/4}(2x/\tau) \tag{1.20}$$

This solution is obtained for an initial mass m distributed on a ring of radius r_0 with a delta Dirac function $\delta(r - r_0)$

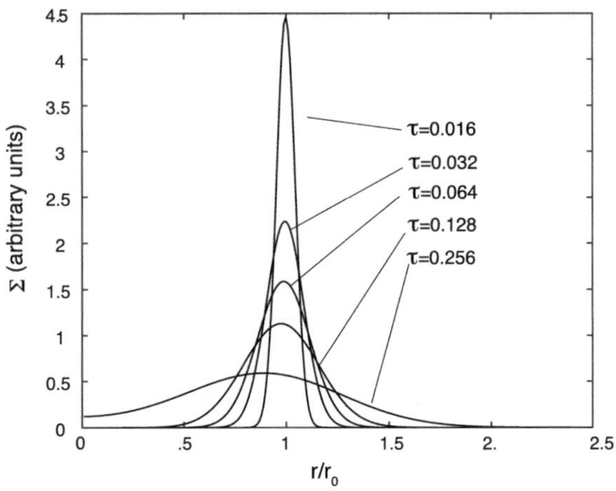

Fig. 1.2 The evolution of surface density in the nebula

$$\Sigma(r, t = 0) = \frac{m}{2\pi r_0}\delta(r.r_0) \tag{1.21}$$

in (1.20) $x = r/r_0$ and the non dimensional time $\tau = 12\nu t/r_0^2$, while $I_{1/4}$ is the modified Bessel function. Before we discuss (1.20) we can show a more intuitive solution of (1.15) with a change of variables

$$X = 2r^{1/2}, \qquad f = \frac{3}{2}\Sigma X \tag{1.22}$$

Then how we show in the appendix we have

$$\frac{\partial f}{\partial t} = D\frac{\partial^2 f}{\partial X^2} \tag{1.23}$$

where

$$D = \frac{\nu}{12X^2} \tag{1.24}$$

As a classical diffusion equation (1.22) gives a viscous time due of the order of ν/r^2 which results to be of a few million years from recent observation of protoplanetary disk. The solution (1.20) has been obtained for an initial condition of a Dirac delta and in **Fig.** 1.2 it is shown the evolution of the non dimensional quantity $\pi(r_0\Sigma/m)$ as a function of a non dimensional time τ. It is interesting at this point to notice that we could define a *viscous time* as

$$t_\nu = \frac{r^2}{\nu} \tag{1.25}$$

Then we can compare this time with the dynamic time scale $(1/\Omega)$

$$\frac{t_\nu}{t_{dyn}} = \frac{r^2\Omega}{\nu} \tag{1.26}$$

To express this ratio in measurable quantities we notice that the mean free path $\lambda = 1/n\sigma$ where n is the number density of the gas and σ the cross section. Then if μ is the average molecular weight

$$\lambda = \frac{\mu m_p}{\rho\sigma} = \left(\frac{\mu m_p}{\sigma\Sigma}\right) H \tag{1.27}$$

where the density ρ has been written as Σ/H where H is an indication of the thickness of the disk. This result could give us some idea about the viscosity. The estimation for Σ follow the assumption of a mass for the disk $\approx 5 \times 10^{-3} M_\odot$ and a radius of ≈ 50 A.U. so we get $\Sigma \approx 10\,\text{g cm}^{-2}$ we also have $\nu = C_c\lambda$ and also assume that $c_s \approx \Omega H$. After a few manipulation we get

$$\frac{t_\nu}{t_{dyn}} = \left(\frac{\sigma\Sigma}{\mu m_p}\right)\left(\frac{r}{H}\right)^2 \tag{1.28}$$

Using some typical value like $\sigma \approx 10^{-16}\,\text{cm}^2$, $m_p \approx 10^{-24}$ g and $H/r \approx 0.1$ we get

$$\frac{t_\nu}{t_{dyn}} \approx 10^{11} \tag{1.29}$$

which mean that based on a dynamical timescale of a few years it would take more than the age of the universe ($\approx$13 *billion years*) to assembly a disk. The question of viscosity will be reexamined later after we learn something about the structure of the disc.

1.3 Radial and Vertical Equilibrium

We will start with an indication why an uniform sphere of radius r under the action of gravity and centrifugal forces flattens. The ratio between these two forces is given by

$$\frac{\Omega^2 r}{Gm(r)/r^2} = \left[\frac{\Omega^2 r^4}{GM(r)}\right]\frac{1}{r} \tag{1.30}$$

So that while the sphere the angular momentum (Ωr^2) stays constant and the ratio will increase until the flattening normal to the rotation axis is halted. However the gravity parallel to the axis will stay constant and the cloud will flatten. We now consider the particular case of a thin disk when the vertical extension is small with respect the horizontal dimension and also assume that the surface density decreases linearly with radius r that is $\Sigma = \Sigma_0 r_0/r$. Then we consider a ring of radius r and

Fig. 1.3 The geometry of the vertical acceleration in the nebula

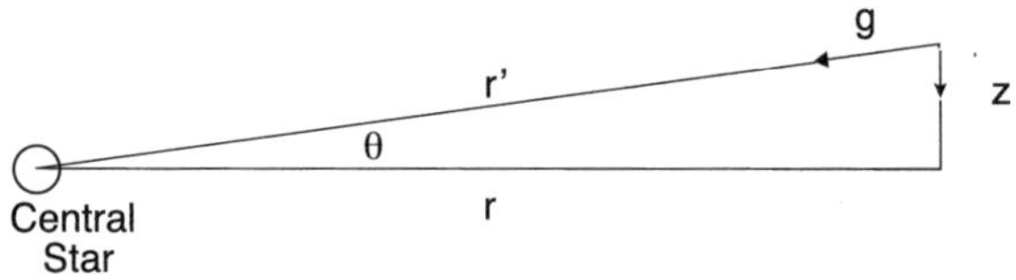

width dr such that its mass is given by $dm = 2\pi r \Sigma(r) dr$ then we have the gradient of the gravitational potential (the acceleration) at the disc midplane

$$\frac{d\Phi}{dr} = 2\pi G \Sigma(r) \tag{1.31}$$

If v_ϕ is the azimuthal velocity the equilibrium between gravity and centrifugal force dictates $v_\phi^2 = r d\Phi/dr$ so that

$$v_\phi^2 = 2\pi G \Sigma(r) r = 2\pi G \Sigma_0 r_0 \tag{1.32}$$

And as consequence the angular velocity $\Omega = v_\phi/r$ will decreases with the increasing radius $\Omega \propto 1/r$. This is in case the gravitational contribution of the disc dominates. On the other hand when the single central object of mass M prevails then the azimuthal velocity is simply $\sqrt{GM/r}$. We notice now that the mass of the disc of radius r

$$M_{\text{disc}} = \int_0^r 2\pi \Sigma r' dr' = \int_0^r 2\pi \Sigma_0 r_0 dr' = 2\pi \Sigma_0 r_0 r = 2\pi \Sigma(r) r^2 \tag{1.33}$$

The field of a single object of mass M is simply GM/r^2 that compared with (1.31) gives $M_{\text{disc}} = M$ that is to produce the same effect the mass of the disc must be comparable with that of the central object. The fact that the inner region of the disc rotates with a higher angular velocity creates a shear which will be crucial in explaining the viscous stress within the disc. As for the vertical structure we assume hydrostatic equilibrium in thew vertical direction z. If P is the pressure and ρ density we have

$$-\frac{1}{\rho}\frac{\partial P}{\partial z} = \frac{\partial \Phi}{\partial z} \tag{1.34}$$

Now if we consider the distance from the center as r' at the vertical distance z we need to consider the vertical component of gravity (see **Fig.** 1.3) as

$$-\frac{GM}{r'^2}\sin\theta = -\frac{GM}{r'^2}\frac{z}{r'} \approx -\frac{GM}{r^2}\frac{z}{r} \tag{1.35}$$

We can the re write (1.34) as

$$\frac{1}{\rho}\frac{\partial P}{\partial \rho}\frac{\partial \rho}{\partial z} = \frac{c_s^2}{\rho}\frac{\partial \rho}{\partial z} = \frac{GM}{r^3}z \tag{1.36}$$

where we have used the definition of sound velocity c_s

$$c_s^2 = \frac{\partial P}{\partial \rho} \tag{1.37}$$

So in case c_s is independent from z (1.36) can be easily integrated

$$\rho(z) = \rho_0 \exp\left[-\frac{GMz^2}{2r^3c_s^2}\right] = \rho_0 \exp\left[-\frac{z^2}{2H_{ng}^2}\right] \tag{1.38}$$

We have introduced a scale-length H_{ng}

$$H_{ng} = \frac{c_s}{\Omega_k} = \frac{c_s r}{v_k} \tag{1.39}$$

This relation shows that if the disc has to be thin $H_{ng} \ll r$ then the azimuthal velocity must be supersonic. We can also calculate the surface density as $\Sigma = \int_{-\infty}^{\infty} \rho(z)dz$ which turns out to be $2\rho_0 H^*$ with $H^* = \sqrt{\pi/2}H_{ng}$. We now turns out to a disc dominated by gravity. In this case the gravitational force has to be replaced by something like (1.31) so we have

$$\frac{c_s^2}{\rho}\frac{\partial \rho}{\partial z} = -2\pi G\Sigma(z) \tag{1.40}$$

In this case because $\Sigma(z) = \int_{-z}^{z} \rho(z')dz'$ the solution is not straightforward (see appendix) but it can be shown that $\Sigma = 2\rho_0 H_{sg}$ where the self gravitating thickness $H_{sg} = c_s^2/(\pi G\Sigma)$. Now we ask how massive must be the disc in order to affect the vertical structure. From (1.35) we know that the z component of gravity for the central star is of the order $(GM/r^2)(H_{ng}/r)$ while that produced by a disc is $2\pi G\Sigma \approx GM_{\rm disc}/r^2$ so that

$$\frac{M_{\rm disc}}{M} \approx \frac{H_{ng}}{r} \ll 1 \tag{1.41}$$

Which means that for a thin disc the vertical structure is affected by the disc gravity even when the mass of the disc is negligible with respect to central mass. If we now consider the ratio

$$\frac{H_{sg}}{H_{ng}} = \frac{c_s\Omega_k}{\pi G\Sigma} \tag{1.42}$$

This expression is equivalent to (1.41) when the factor on the right of (1.42) is of the order of unity. This is related as we will see in the next paragraph to the so called *gravitational instability* which generate the self gravitating disc.

1.4 Transport of Angular Momentum

We started this chapter by acknowledging that most of the mass in the solar system resides in the Sun so we need to justify how angular momentum is exported from the center to the rim. We have then to find a mechanism for the transport of angular momentum. Since the epoch of Lynden Bell (mid seventies) it was though that the simplest mechanism was the dragging due to viscosity of the outer layer (which were slower) from the inner layers (which were faster). The problem is that the viscosity defined by (1.25) would results in too long a time, so people recur to a turbulent viscosity and with an analogy to the mean free path define it as

$$\nu = \alpha c_s H \tag{1.43}$$

where α is a factor much less than 1. The generation of what is called *turbulent viscosity* is attributed to the insurgence of a gravitational instability in the disc. This kind of instability arises when in a uniform disc of gas and dust due to turbulent motions some clump of higher density may form so to attracts more matter. This instability may lead either to the formation of spiral galaxy or to planets. The conditions for the instability can be found by considering a clump of dimension Δr and mass $m \approx \pi(\Delta r)^2\Sigma$. If isolated such clump would collapse on the free-fall time scale

$$t_{ff} \approx \sqrt{\frac{\Delta r^3}{Gm}} \approx \sqrt{\frac{\Delta r}{\pi G \Sigma}} \tag{1.44}$$

Stabilizing influences could be pressure and shear. The typical times associated are for pressure the time it takes for a sound wave to traverse the clump

$$t_p \approx \frac{\Delta r}{c_s} \tag{1.45}$$

while for shear we have

$$t_{\text{shear}} \approx \frac{1}{r}\left(\frac{d\Omega}{dr}\right)^{-1} \approx \frac{1}{\Omega} \tag{1.46}$$

We see that t_p is linear with the size of the clump while t_{shear} is not, that means that pressure tends to stabilize small regions while the shear will influence large ones. The disc will be gravitationally unstable to clump formation if t_{ff} will be shorter than either t_p or t_{shear}. Setting the three times equal we have a condition for instability

$$\pi G\Sigma \geq c_s\Omega \tag{1.47}$$

In astrophysics this is mentioned as the *Q Toomre parameter* from the astrophysicist Alar Toomre which formally derived the criteria

$$Q = \frac{c_s\Omega}{\pi G\Sigma} \tag{1.48}$$

Now we have derived this expression as (1.42) that can have a different interpretation considering that $H_{ng} \approx c_s/\Omega$ and $M_{\rm disc} = \pi r^2\Sigma$ we have

$$\frac{H_{ng}}{r} = \frac{c_s}{\Omega r} \approx \frac{\pi G\Sigma}{\Omega^2 r} \approx \frac{\pi r^2\Sigma}{M} \approx \frac{M_{\rm disc}}{M} \ll 1 \tag{1.49}$$

where we have used $Q = 1$ and $\Omega^2 = GM/r^3$ with M central mass which is the same than (1.41). We now introduce the continuity equation in slight different form

$$\dot{M} = -2\pi r v_r \Sigma \tag{1.50}$$

In such way that the accretion rate $\dot{M}$ is positive when v_r is negative (directed inward). In the same way the angular momentum

$$\dot{M}\Omega r^2 - 2\pi\nu\Sigma r^3\frac{d\Omega}{dr} = \dot{J} \tag{1.51}$$

where $\dot{J}$ is the change in angular momentum resulting from two contribution, the first on the left hand side due to advection and the second due to the outward flux produced by the viscous stresses. The determination of $\dot{J}$ is based on the so called *no torque condition*. At some inner radius r_{in} near the central object the presence of a boundary layer imposes the gradient $d\Omega/dr = 0$ so that we have $\dot{J} = \dot{M}(\Omega r^2)_{in} = \dot{M}\sqrt{GMr_{in}}$. If we substitute in (1.51) considering that

$$\frac{d\Omega}{dr} = \frac{3}{2}\frac{\Omega}{r}$$

we find easily that

$$3\pi\nu\Sigma = \dot{M}\sqrt{1 - \frac{r_{in}}{r}} \tag{1.52}$$

And at large distances $r \gg r_{in}$ we have

$$\dot{M} = 3\pi\nu\Sigma = 2\pi\nu\Sigma\left|\frac{d\ln\Omega}{d\ln r}\right| \tag{1.53}$$

In case we use (1.43) where $H = c_s^2/G\Sigma$ we have

$$\dot{M} = 2\alpha \frac{c_s^2}{G} \left| \frac{d \ln \Omega}{d \ln r} \right| \tag{1.54}$$

And the accretion rate only depends on the speed of sound.

1.5 Temperature Profile

We now would like to inquire about the temperature in the disc as a function of radius r. The hypothesis is that the disk is heated by the energy dissipated by viscosity. It can be shown (see appendix) that the energy dissipated per unit time and unit area $D(r)$ is given at large distance by

$$D(r) = \tfrac{1}{2} \nu \Sigma \left(r \frac{d\Omega}{dr} \right)^2 \tag{1.55}$$

When substituting (1.52) we have

$$D(r) = D(r) = \tfrac{1}{2} \frac{\dot{M}}{3\pi} \left(r \frac{d\Omega}{dr} \right)^2 = \frac{3GM\dot{M}}{8\pi r^3} \tag{1.56}$$

where we have used for Ω the keplerian value. To find the temperature T_e we assume that this energy is dissipated in radiation at the rate $2\sigma T_e^4$. The factor of two is because the disc emits from both sides. We then find

$$T_e = \left(\frac{3GM\dot{M}}{8\pi \sigma r^3} \right)^{1/4} \tag{1.57}$$

The relevant fact is that temperature is a function of r and then different regions of the disc will emit radiation of different characteristics so that the spectrum of radiation will be a quite complicated matter.

1.6 Let Us Now Give Some Numbers

During this long excursus we did not give any numbers on the different quantities. Now we would like to know if the relations we got can be compared with the scarce data we have on the primitive nebula. We will make some order on the various quantities.

(a) Sound Speed We may refer to the fact that the gas in the nebula is molecular hydrogen and for it we use a an isothermal transformation so that we have

$$p = \rho k T \tag{1.58}$$

where k is the Boltzmann constant, ρ is the density and T the temperature. From the definition of speed of sound $c_s = \sqrt{dp/d\rho}$ we get

$$c_s = \sqrt{\frac{p}{\rho}} = \sqrt{\frac{kT}{\mu m_H}} \tag{1.59}$$

with μ molecular weight and m_H mass of the hydrogen atom. In case of an adiabatic transformation the relation between density and temperature is given by considering that $pV^{5/3} = cost$ and that density goes like $1/V$

$$p\rho^{-5/3} = cost \tag{1.60}$$

so that the sound velocity

$$c_s = \left(\frac{5}{3}\frac{p}{\rho}\right)^{1/2} = \sqrt{\frac{5}{3}\frac{kT}{\mu m_H}} \tag{1.61}$$

which is of the same order of (1.59) that for a temperature of 300 K gives $c_s \approx$ 1100 m/s.

(b) Viscosity For viscosity we use (1.43) where then scale height can be found from (1.49). The for 1 A.U. $\approx 1.5 \times 10^{11}$ m and $M_{\text{disc}}/M \approx 0.1$ m we get $H \approx 0.5 \times 10^{10}$ m. Then if we use $\alpha \approx 0.01$ we get $\nu \approx 1.5 \times 10^{11}\,\text{m}^2\,\text{s}^{-1}$, This is a quite large value for the viscosity.

(c) Temperature We may consider a more general relation than (1.57) valid also for small distances from the central body of radius r^* that reads

$$T_e = \left\{\frac{3GM\dot{M}}{8\pi\sigma r^3}[1 - (r^*/r)^{1/2}]\right\}^{1/4} \tag{1.62}$$

Now we have plotted the results of (1.62) using the following data, $G = 6.67 \times 10^{-11}\,\text{m}^3\,\text{kg}^{-1}\,\text{s}^{-2}$, $M = 1.98 \times 10^{30}$ kg, $\dot{M} = 2 \times 10^{-5} M_{\odot}/year$. **Figure** 1.4 shows the results together with those of a numerical model of Chiang and Goldreich. Actually the accretion rate we have used may be still too high so that these results would pointing out the necessity for an additional source of energy.

Appendix

Equation of Continuity

We consider an annulus extending between r and $r + \Delta r$ and surface density Σ we have for the conservation of mass

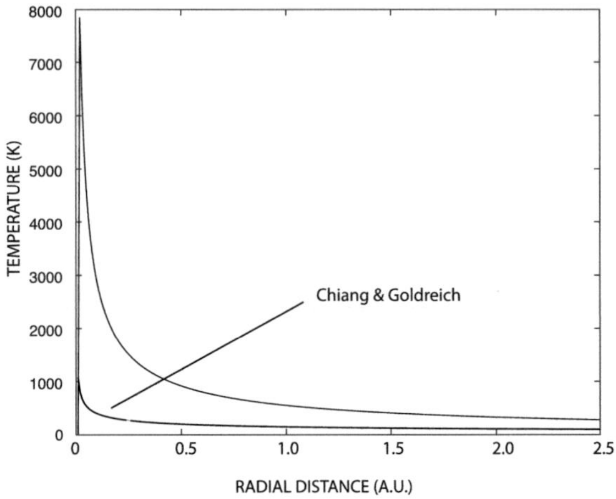

Fig. 1.4 The temperature profile in the nebula compared to the calculations of Chiang and Goldreich

$$\frac{\partial}{\partial t}(2\pi r \Delta r \Sigma) = 2\pi r \Sigma(r) v_r(r) - 2\pi (r + \Delta r)\Sigma(r + \Delta r) v_r(r + \Delta r) \quad (1.63)$$

where we write $\Sigma(r + \Delta r) = \Sigma(r) + (\partial \Sigma / \partial r)\Delta r$ and taking the limit for $\Delta r \to 0$

$$\frac{\partial \Sigma}{\partial t} + \frac{1}{r}\frac{\partial}{\partial r}(r \Sigma v_r) = 0 \quad (1.64)$$

This equation implies no changes in the azimuthal and vertical velocities. In the most general case the equation of continuity in cylindrical coordinate reads

$$\frac{\partial \Sigma}{\partial t} + \frac{1}{r}\frac{\partial}{\partial r}(r \Sigma v_r) + \frac{1}{r}\frac{\partial}{\partial \phi}(\Sigma v_\phi) + \frac{\partial}{\partial z}(\Sigma u_z) = 0 \quad (1.65)$$

The Diffusion Equation

We will from Eq. (1.15) when $\nu = \text{cost}$.

$$\frac{\partial \Sigma}{\partial t} = \frac{3\nu}{r}\frac{\partial}{\partial r}\left[r^{1/2}\frac{\partial}{\partial r}(\Sigma r^{1/2})\right]$$

and make the substitution

$$X = 2r^{1/2}, \qquad f = \frac{3}{2}\Sigma X, \qquad r = \frac{X^2}{4}$$

Then use the rule

$$\frac{\partial}{\partial r} = \frac{\partial X}{\partial r}\frac{\partial}{\partial X} = \frac{2}{X}\frac{\partial}{\partial X}$$

Then we notice that

$$\Sigma r^{1/2} = \frac{f}{3} \rightarrow \frac{\partial \Sigma}{\partial t} = \frac{2}{3X}\frac{\partial f}{\partial t}$$

Making all the substitution we get

$$\frac{\partial f}{\partial t} = D\frac{\partial^2 f}{\partial X^2}, \qquad D = \frac{12\nu}{X^2} \tag{1.66}$$

The Solution for the Surface Density

We will show how to solve Eq. (1.66) that can written

$$\frac{\partial}{\partial t}(r^{1/2}\Sigma) = \frac{12\nu}{X^2}\frac{\partial^2}{\partial X^2}(r^{1/2}\Sigma) \tag{1.67}$$

We can separate the variables $r^{1/2}\Sigma = T(t)S(X)$ so that

$$\frac{1}{T}\frac{dT}{dt} = \frac{12\nu}{X^2}\frac{1}{S}\frac{d^2 S}{dX^2} = const = -\lambda^2$$

The function $T(t)$ is the an exponential while the function $S(X)$ is a Bessel function. We use an initial condition a Dirac Delta function with a ring of mass m at $r = r_0$

$$\Sigma(r, t = 0) = \frac{m}{2\pi r_0}\delta(r - r_0)$$

With the further substitution $x = r/r_0$ and $\tau = 12\nu t r_0^{-2}$ we get the solution

$$\Sigma(x, \tau) = \frac{m}{\pi r_0^2}\frac{1}{\tau x^{1/4}}\exp\left[-\frac{(1 + x^2)}{\tau}\right] I_{1/4}(2x/\tau)$$

where $I_{1/4}$ is the modified Bessel function.

Self Gravitating Disc

We consider the most general case when we solve both the hydrostatic equilibrium and the Poisson equation

$$\frac{1}{\rho}\nabla p = -\nabla\Psi, \qquad \nabla^2\Psi = 4\pi G\rho \tag{1.68}$$

where $p = c_s^2\rho$ so that the first in (1.68) becomes

$$\frac{c_s^2}{\rho}\nabla\rho = -\nabla\Psi \tag{1.69}$$

That can be specialized over z to give

$$c_s^2\frac{d}{dz}(\ln\rho) = -\frac{d\Psi}{dz} \tag{1.70}$$

That can be integrated to give

$$\rho = \rho_0 \exp -\frac{\Psi - \Psi_0}{c_s^2} \tag{1.71}$$

where ρ_0 refer to $z = 0$. Substituting in (1.68)

$$\nabla^2\Psi = 4\pi G\rho_0 \exp -\frac{\Psi - \Psi_0}{c_s^2} \tag{1.72}$$

Now we make a change of variable

$$Z = z\frac{\sqrt{2\pi G\rho_0}}{c_s}, \qquad X = -\frac{\Psi - \Psi_0}{c_s} \tag{1.73}$$

The it is easy to show that (1.72) becomes

$$\frac{d^2X}{dZ^2} = -2e^X, \qquad \frac{1}{2}\frac{d}{dZ}\left[\left(\frac{dX}{dZ}\right)^2\right] = -2\frac{d}{dZ}(e^X)$$

Which can be integrated with the condition $dX/dZ = 0$ when $Z = 0$

$$\left(\frac{dX}{dZ}\right)^2 = C - 4e^X, \quad \rightarrow \frac{dX}{dZ} = 2\sqrt{1-e^X}, \quad \rightarrow \int\frac{dX}{\sqrt{1-e^X}} = 2\int dZ$$

The integral can be solved with the substitution $e^X = sin^2\theta$

$$\int\frac{dX}{\sqrt{1-e^X}} = \int\frac{2\cos\theta d\theta}{\sin\theta\sqrt{1-\sin^2\theta}} = \int\frac{2d\theta}{\sin\theta}$$

and with the further substitution $t = \tan\theta/2$ we get

$$\int \frac{2d\theta}{\sin\theta} = \int \frac{1+t^2}{t} d\theta = 2\int \frac{dt}{t} = 2\ln t + D$$

And with the condition $X = 0$ for $Z = 0$ we get D=0. And the Poisson equation becomes

$$2\ln t = 2Z + D, \rightarrow t = e^Z$$

Now we notice that

$$\sin\theta = 2sin(\theta/2)\cos(\theta/2) = \frac{2\sin\frac{\theta}{2}\cos\frac{\theta}{2}}{\sin^2\frac{\theta}{2} + \cos^2\frac{\theta}{2}} = \frac{2t}{1+t^2} = \frac{2e^Z}{1+e^{2Z}} = \frac{1}{\cosh Z}$$

So that we have

$$\Psi - \Psi_0 = 2c_s^2 \ln\cosh\left(\frac{\sqrt{2\pi G\rho_0}}{c_s} z\right), \quad \rightarrow \rho = \frac{\rho_0}{\cosh^2 \frac{\sqrt{2\pi G\rho_0}}{c_s} z} \tag{1.74}$$

So we have a scale height $H = c_s/\sqrt{2\pi G\rho_0}$. This can be expressed as a function of the surface density Σ

$$\Sigma = \int_{-\infty}^{\infty} \rho dz = c_s\sqrt{\frac{2\rho_0}{\pi G}} = \frac{c_s^2}{\pi G H} = 2\rho_0 H \rightarrow H = \frac{c_s}{\sqrt{2\pi G\rho_0}} \tag{1.75}$$

Dissipation Rate

Remember the expression for the total torque (1.10)

$$\mathrm{T}(r) = 2\pi r\nu\Sigma r^2 \frac{d\Omega}{dr}$$

This refers to torque between two adjacent rings (**Fig.** 1.1). This expression gives the right behavior because the torque is zero in the absence of shear ($d\Omega/dr = 0$) and the toque is negative when ($d\Omega/dr < 0$) that is the angular velocity decreases going outward and the inner rings lose angular momentum and the gases and dust spirals toward the center. The net torque between two adjacent rings can be calculated as

$$\mathrm{T}(r+dr) - \mathrm{T}(r) = \frac{\partial \mathrm{T}}{\partial r} dr \tag{1.76}$$

Then the associated power is given by

$$\Omega \frac{\partial \mathrm{T}}{\partial r} dr \tag{1.77}$$

This power is the loss rate of mechanical energy to dust and gas and represents the *the viscous dissipation* and considering that the disc has two surfaces of a total area $4\pi r dr$ the dissipation rate will be

$$D(r) = \frac{\mathrm{T}}{4\pi r}\frac{d\Omega}{dr} = \frac{1}{2}\nu\Sigma\left(r\frac{d\Omega}{dr}\right)^2 \tag{1.78}$$

In case the rotation is keplerian then

$$\Omega = \sqrt{\frac{GM}{r^3}} \tag{1.79}$$

Equation (1.78) becomes

$$D(r) = \frac{9}{8}\nu\Sigma\frac{GM}{r^3} \tag{1.80}$$

G in this case is the gravitational constant

A Toy Model for the Nebula

It is nice to mention an old paper by Coradini et al. They considered the disc originated from a sphere of mass $M \approx 1.01 M_{\odot}$ and density ρ_0 with $M_{\odot}$ the central mass that rotates with velocity Ω_0, Then from the conservation of mass the volumetric density of the disc is

$$\rho_0' = \rho_0\frac{M_d}{M_d + M_{\odot}} \tag{1.81}$$

Now we compare the angular momentum of two elementary cylinder as shown in **Fig.** 1.5 where the large cylinder with momentum $d\Psi_N$ changes in the cylinder with momentum $d\Psi_D$ that can be calculated as

$$d\Psi_N = 4\pi\rho_0'\Omega_0 R^5 \sin^3\theta\cos^2\theta d\theta, \qquad d\Psi_D = 2\pi\left(\frac{GM_{\odot}}{r^3}\right)^{1/2}\Sigma r^2 dr \tag{1.82}$$

Imposing conservation of angular momentum we have the surface density

$$\Sigma = \frac{2\rho_0'}{r^2}\Omega_0 R^5 \sin^3\theta\cos^2\theta\left(\frac{GM_{\odot}}{r^3}\right)^{-1/2}\frac{d\theta}{dr} \tag{1.83}$$

The conservation of angular momentum for a single particle at the edge of the disc dictates

$$\Omega_0 R^2 \sin^2\theta = \{GM_{\odot}r\}^{1/2} \tag{1.84}$$

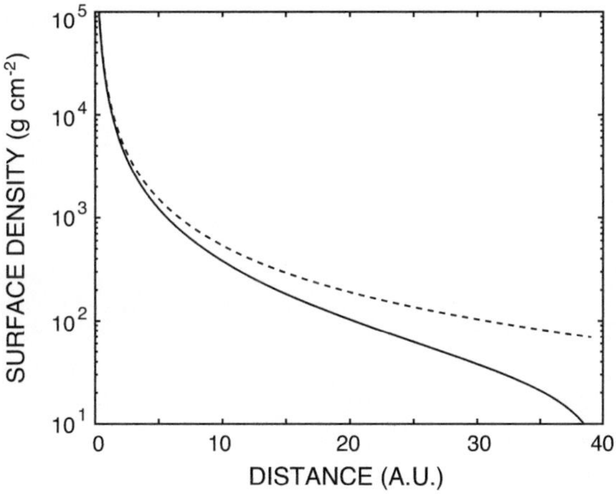

Fig. 1.5 The surface density as a function of the distance from the center in astronomical units. The dashed line refers to the $r^{-3/2}$ approximation

From which we can calculate $d\theta/dr$ that substituted into (1.83) gives

$$\Sigma = \frac{\rho_0 R}{2\Omega_0}\left(\frac{GM_\odot}{r^3}\right)^{1/2}\left[1 - \frac{(GM_\odot r)^{1/2}}{\Omega_0 R^2}\right]^{1/2} \tag{1.85}$$

We now assume $\rho_0 \approx 10^{-16}$ kg/m^3 and from (1.81) get $\rho_0' \approx 10^{-18}$ kg/m^3 and we can compute R

$$R = \left[\frac{(M_d + M_\odot}{(4/3)\pi\rho_0}\right]^{1/3} \approx 1.7 \times 10^{15} \text{ m}$$

The assuming $r_d \approx 40$ A.U. we can get Ω_0 from (1.84) taking $\sin\theta \approx 1$

$$\Omega_0 = (GM_\odot r_d)^{1/2} R^{-2} \approx 10^{-14}\, s^{-1} \approx 3 \times 10^6\, years$$

So we can now explicate the surface density valid up to a distance 0f 40 A.U.

$$\Sigma = 1.682 \times 10^4 r_{au}^{-3/2}[(1 - 0.158 r_{au}^{1/2})]^{1/2} \tag{1.86}$$

where r_{au} is the radius expressed in astronomical units. This result it is shown in **Fig.** 1.5 together with some other hypothesis. At this point to find the temperature in the disc we proceed in the same way as we did before just change the dissipation rate. Using (1.80) we calculate the dissipated power because we want to take into account the radial dependence of the surface density

$$D(r) = \frac{9}{8}\nu\Sigma_0 r^{-3/2}\frac{GM}{r^3} \tag{1.87}$$

where we have taken only the dependence $^{-3/2}$ and $\Sigma_0 = 1.682 \times 10^4$. The so called *luminosity* is the total power emitted

$$L(r) = \frac{9}{8} G M \pi \nu \int_{r^*}^{r} 1.68 \times 10^4 r^{-3/2} r^{-3} dr \tag{1.88}$$

where r^* is the inner radius of the boundary layer, We get finally

$$L(r) = \frac{9}{10} \pi \nu \Sigma_0 G M \left(r^{-5/2} - r*^{-5/2} \right) \tag{1.89}$$

References

Textbooks

Armitage, P.J.: Astrophysics of Planet Formation. Cambridge University Press (2010)
Armitage, P.J., Kley, W.: From Protoplanetary Disc to Planet Formation. Springer (2019)
Coradini, A., Federico, C., Magni, C.: Some remarks on the formation of terrestrial planets. Comparative Study of the Planets, NATO Advanced Course. Springer (1981)
Frank, J., King, A., Raine, D.: Accretion Power in Astrophysics. Cambridge University Press (2002)
Lodders, K., Fegley, B.: Royal Chemistry Society, Chemistry of the Solar System (2010)

Articles

Armitage, P.J.: Dynamics of protoplanetary disks. Ann. Rev. Astron. Astrophys. **49**, 195 (2011)
Kratter, K.M., Lodato, G.: Gravitational instabilities in circumstellar disks. Ann. Rev. Astron. Astrophys. **54**, 271–311 (2016)
Lodato, G.: Self-gravitating accretion discs. Nuovo Cimento Rivista Serie **30**, 293 (2007)
Lynden-Bell, D., Pringle, J.E.: The evolution of viscous discs and the origin of nebular variables. Mon. Not. R. Astr. Soc. **168**, 603 (1974)
Pringle, J.E.: Accretion discs in astrophysics. Ann. Rev. Astron. Astrophys. **19**, 137 (1981)
Toomre, A.: On the gravitational stability of a disc of stars. Ap. J. **139**, 1217 (1964)
Weidenschilling, S.J.: The distribution of mass in the planetary system and solar nebula. Astrophys. Space Sci. **51**, 153 (1977)

Chapter 2
The Formation of the Earth

We have now some information on the primordial disc. These objects are now being observed by Earth based and spaceborne telescopes. In this chapter we would like to study how from this disk the terrestrial planets (and so the Earth) may form. The key to understand this process is the *gravitational instability* of a flat object. We should be familiar with the Jeans instability which refers to a limit condition when on a spherical object the pull of gravity overcome the pressure force while we are much less familiar when the instability refer to a flat object. The pieces resulting from this instability may give origin to the embryo of the planet that we call *planetesimals*. The subsequent interactions of such planetesimals give raises to rocky planets. To illustrate this path we will follow the illuminating work of Philip Armitage while this mechanism was first introduced by Viktor Safronov of the Moscow University in a precious little book of 1969 and by Peter Goldreich and Robert Ward of Caltech in a paper five years later.

2.1 The Gravitational Instability and Planetesimals Size

If you look around for a physical explanation of the gravitational instability for a flat object you do not find any. Most of the people (as we will do in the appendix) just solve the continuity and momentum equation with a wave solution and look for the amplification condition. So the most complete definition is still that of Safronov: *A medium is gravitational unstable if new developed density perturbations in it, however small, increase indefinitely with time due to gravity and disrupt the equilibrium.* We have already mentioned the instability criteria in the paragraph 1.4 of previous chapter as the inequality

$$\pi G \Sigma \geq c_s \Omega$$

G. Visconti, *Origin and Evolution of the Atmosphere and Ocean*,
https://doi.org/10.1007/978-3-032-02653-8_2

where G is the gravitational constant, Σ is the surface density, c_s is the sound speed and Ω is the rotation rate. This criteria translate in the so called *Toomre parameter*

$$Q = \frac{c_s \Omega}{\pi G \Sigma} \leq 1$$

Now we consider that once the instability has occurred the fragments have a velocity dispersion σ to which we associate a vertical scale height

$$h_d = \frac{\sigma}{\Omega} \tag{2.1}$$

To evaluate the most unstable length of the perturbation we notice that $c_s \Omega$ has the dimension of an acceleration and we can form a length

$$\lambda = \frac{2\sigma^2}{G\Sigma} \tag{2.2}$$

Which implies that the mass of the fragment is of the order

$$m_p \approx \pi \lambda^2 \Sigma \approx 4\pi \frac{c_s^2}{G^2 \Sigma} \approx 4\pi^5 G^2 Q^4 \frac{\Sigma^3}{\Omega^4} \tag{2.3}$$

where we have assumed that $\sigma \approx c_s$ and have used the definition of Q to eliminate c_s. Using the values at the Earth orbit $\Sigma = 10$ g cm^{-2}, $\Omega = 2 \times 10^{-7}$ s^{-1} and taking $Q = 1$ we get

$$m_p(1AU) \approx 3 \times 10^{15} \text{ kg} \tag{2.4}$$

which corresponds to a sphere of 5–10 km in radius.

A necessary condition for the instability to arise is that gas and particles of the primitive nebula to be confined in a very thin disc. This happens because solid particles tend to settle in the midplane of the disc and due to a radial drift migrate toward the central mass of the nebula. The settling particle is subject to two forces, the vertical gravity given by (1.35)

$$F_{grav} = m\Omega_k^2 z \tag{2.5}$$

And the aerodynamic drag force

$$F_D = \frac{4}{3}\pi a^2 \bar{v} \rho v \tag{2.6}$$

Which refer to a particular hydrodynamic regime with $\bar{v}$ the thermal speed of the molecules. To find the settling velocity we equate the last two expressions and use the mass $= 4\pi \rho_d a^3/3$ with ρ_d density of the particle to get

$$v_{set} = \left(\frac{\Omega_k^2}{\bar{v}}\right)\frac{\rho_d}{\rho}az \quad (2.7)$$

As an indication for micron sized particles at an altitude $z \approx h \approx 0.04$ AU we obtain $v_{set} \approx 0.1$ cm s^{-1} which corresponds to settling time of

$$t_{set} = f\frac{h}{v_{set}} \approx 2 \times 10^5 years \quad (2.8)$$

Which is short with respect the lifetime of the disc. Another possible effect is related to the presence of diffusion which should inhibit the layering of the particles. As we know the diffusion time should look like

$$t_{diff} \approx \frac{z^2}{D} \quad (2.9)$$

where the diffusion coefficient D may be taken as

$$D \approx \nu = \frac{\alpha c_s}{\Omega_k} \quad (2.10)$$

It can be shown that the minimum value of α for which turbulence (i.e. diffusion) will oppose sedimentation is given by

$$\alpha \geq \frac{\pi}{2}e^{1/2}\frac{\rho_d a}{\Sigma} \quad (2.11)$$

This condition gives for $a = 1$ μm, $\alpha \geq 10^{-5}$ which means that most small particles will remain suspended. If a particle move toward the midplane of the disc it may capture other particles along its path and grow by coagulation. If the fall happens with velocity v_{set} it will swept a volume in the unit time given by $\pi a^2 v_{set}$ so that the mass growth will be

$$\frac{dm}{dt} = \pi a^2 v_{set} f\rho(z) \quad (2.12)$$

where f is the dust to gas ratio in the disk. If we substitute (2.7) we get

$$\frac{dm}{dt} = \frac{3}{4}\frac{\Omega^2 f}{\bar{v}}zm \quad (2.13)$$

To find the solution $m(t)$ we must solve also (2.7)

$$\frac{dz}{dt} = v_{set} = \frac{\Omega^2}{\bar{v}}\frac{\rho_d}{\rho}az \quad (2.14)$$

We do this exercise in the appendix and the results are shown in **Fig.** 2.1. It can be seen that the time scale to settle in the midplane is of the order of 1000 years. In

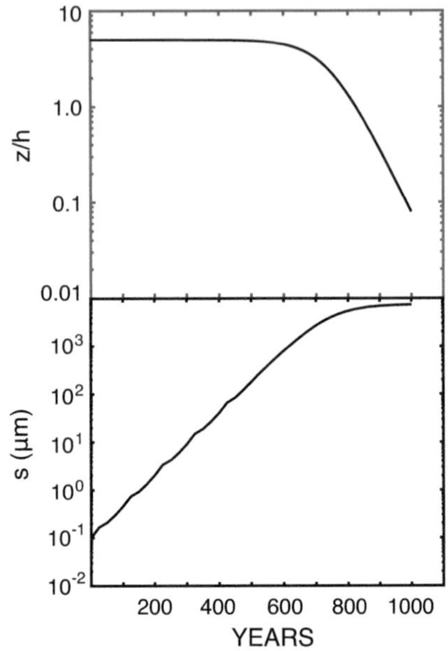

Fig. 2.1 The eight above the midplane and the radius of an accreting particle by sedimentation

addition to the settling motion there is a movement of the particle toward the center, known as *radial drift*. This consists in a motion that for small particle ($a > 1$ cm) is due to the fact that they do not experience the same pressure force than the gas and so fall inward. For larger particles ($a > 1$ m) they move at the keplerian velocity faster than the gas and experience an aerodynamic drag that make them to fall inward. The radial equilibrium demand

$$\frac{v_{\phi,\mathrm{gas}}}{r^2} = \frac{GM}{r^2} + \frac{1}{\rho}\frac{\partial P}{\partial r} \tag{2.15}$$

Then assume a radial dependence of pressure like

$$P = P_0 \left(\frac{r}{r_0}\right)^{-n} \tag{2.16}$$

That substituted in (2.15) gives

$$v_{\phi,\mathrm{gas}} = v_k \sqrt{1 - n\frac{c_s^2}{v_k^2}} \tag{2.17}$$

where $v_k = \Omega_k r$ so that the azimutal velocity for the gas is always smaller then the keplerian velocity. It easy to find that at 1 A.U. for typical values of n and Σ we have $v_{\phi,\mathrm{gas}} = 0.996 v_k$ and still the solid particles will experience a considerable drag. We can go back to (2.6) and define a *friction time*

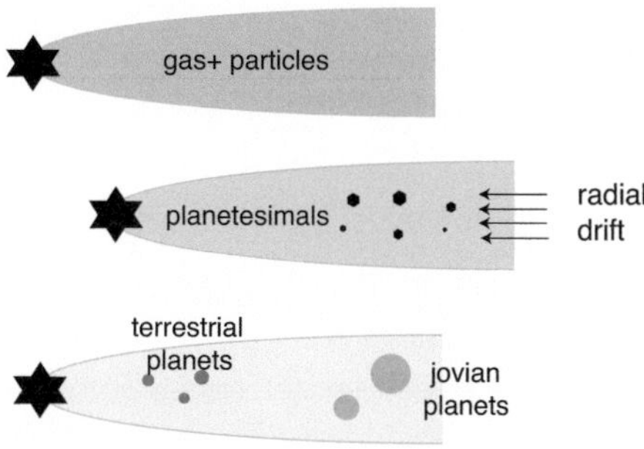

Fig. 2.2 The evolution of the protoplanetary disc. The initial distribution evolves in a thinner region to the formation of planetesimal confined in the midplane. Adapted from Armitage

$$t_{\text{frict}} = \frac{mv}{F_D} = \frac{\rho_d}{\rho}\frac{a}{\bar{v}} \tag{2.18}$$

For the radial drift resulting from the difference between $v_{\phi,\text{gas}}$ and v_k we can introduce another dimensionless time constant as

$$\tau_{\text{frict}} = t_{\text{frict}}\Omega_k \tag{2.19}$$

such that

$$\frac{v_r}{v_k} = -\frac{\eta\tau_{\text{frict}}}{1+\tau_{\text{frict}}^2} \tag{2.20}$$

With $\eta = n(c_s/v_k)^2$. Notice that because $h = c_s/\Omega_k$ then $c_s/v_k = h/r$ and that the maximum ratio occurs for $\tau_{\text{frict}} \approx 1$. The overall results is that $v_r \ll v_k$ so that the particle will spiral toward the central mass and the maximum in spiral rate is for particle sizes between 10 cm to several meters while the lifetime of the particle at 1 AU is of the order of 100 years much less than the lifetime of the disk. The net result of the dynamics within the disc is motion of particles toward the midplane and a spiraling toward the center. The simplest evolution is illustrated in **Fig.** 2.2 with the planetesimals mostly confined in the midplane with a mass distribution obeying a power law like

$$\frac{dN}{dM_p} \propto M_p^{-p} \tag{2.21}$$

We will see in the next paragraph how this population of planetesimals can originate the terrestrial planets.

2.2 The Growth to Terrestrial Planets

We assume that at this point planetesimals have formed in the protoplanetary disc. These objects have dimension of a few km and their gravity is large enough to perturb the orbit of similar objects. Collisions with other planetesimals rapid growth

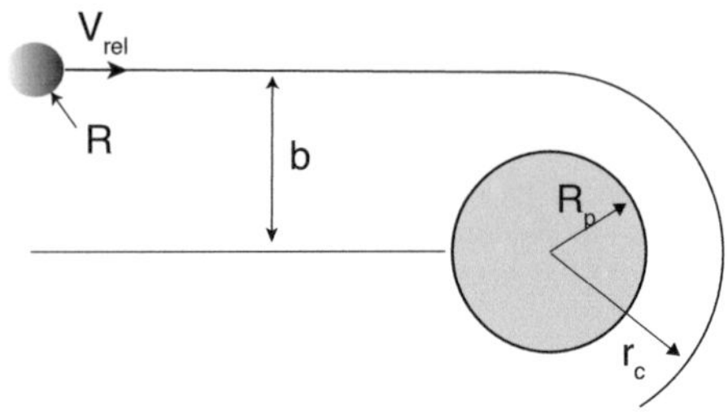

Fig. 2.3 The collision between a large object of radius R_p with an incoming object of radius R

to *protoplanets*. These have dimension similar to the Moon or Mars (a couple of thousands km) and when they have formed the disc is depleted completely of gas. Further interactions between these protoplanets are responsible for the formation of terrestrial planets. We begin with some recall of celestial mechanics. One of the mechanism which increases the probability of collision is the *gravitational focusing*. Consider as in **Fig.** 2.3 a growing planetesimal of radius R_p and the accreting object of radius R with $R \ll R_p$. The geometric cross section is $\sigma_g = \pi R_p^2$ however if we consider the effects of gravity the effective cross section is much larger. We assign to the particle mass m and an *impact parameter* b. Also the velocity at large distance is $v_{\rm rel}$. And we define r_c the distance of closest approach and m_p the mass of the planetesimal. Then we apply the conservation of momentum and energy

$$\begin{aligned} b v_{\rm rel} &= R_p v_c \\ \frac{1}{2}\mu v_{\rm rel}^2 &= \frac{1}{2}\mu v_c^2 - \frac{G\mu M_t}{R_p} \end{aligned} \tag{2.22}$$

where $\mu = m m_p/(m + m_p)$ is the reduced mass and $M_t = m + m_p$ while v_c is the velocity at the closest approach. We obtain v_c from the conservation of momentum and substitute in the second to get

$$\left(\frac{b}{R_p}\right)^2 = 1 + \frac{2GM_t}{R_p v_{\rm rel}^2}$$

We now introduce a *joint escape velocity*

$$v_{esc} = \left(\frac{2GM_t}{R_p}\right)^{1/2} \tag{2.23}$$

To get the new cross section

$$\pi b^2 = \pi R_p^2 F_g = \pi R_p^2 \left[1 + \left(\frac{v_{esc}}{v_{\rm rel}}\right)^2\right] \tag{2.24}$$

The focusing factor F_g is clearly important for cold disc when $v_{rek} \ll v_{esc}$ that is slow moving particles will be captured more easily the fast moving particles. This is a classical two body problem but we can imagine in the protoplanetary disc the presence of the central mass and the growing planet. In this case it is important the concept of the *Hill sphere*. We use a dimensional argument by comparing the frequency around the star with mass M_* at distance a with the same frequency for an object of mass M_p at distance r. We get

$$\sqrt{\frac{GM_*}{a^3}} = \sqrt{\frac{GM_p}{r^3}} \tag{2.25}$$

We get the definition of radius of the Hill sphere

$$r_H = \left(\frac{M_p}{M_*}\right)^{1/3} \tag{2.26}$$

The Hill radius is the distance over which the gravity of the planet dominates the gravity of the central star. An example can be made for the Earth (mass $m = 5.97 \times 10^{24}$ kg) and the Sun (mass $M = 1.99 \times 10^{30}$ kg) at a distance of 1.495×10^8 km and the Hill radius computed with more accurate formula

$$r_H = a\left(\frac{m}{3M}\right)^{1/3} \approx 1.495 \times 10^6 \text{ km}$$

This means that the Moon which orbits the Earth at a distance of 3.84×10^5 km will never be captured by the gravity of the Sun and so the Hill radius establish the region dominated by the two body problem.

2.3 Accretion or Disruption

At this point the protoplanets may undergoes to further growth by collisions that could produce an accumulation on the growing planet (*accretion*) or the impact could be resolved in pieces that re-aggregate to a larger body (*shattering*) or rather (*dispersal*) with the dissipation of pieces. An example can be made with two pieces of mass m_1 and m_2. Their ratio will change as

$$\frac{d}{dt}\left(\frac{m_1}{m_2}\right) = \frac{m_1}{m_2}\left(\frac{1}{m_1}\frac{dm_1}{dt} - \frac{1}{m_2}\frac{dm_2}{dt}\right) \tag{2.27}$$

In the case $m_1 > m_2$ we can see that the ratio (m_1/m_2) if the relative grow rate $1/m_i(dm_i/dt)$ increase with the mass. In this case we have a *runaway growth*. On the other hand we have an *ordered growth* if the relative growth decrease with mass.

In this case the ratio m_1/m_2 tends to unity. Based on the cross section (2.24) we can write the mass growth for a planetesimal of mass m_p

$$\frac{dm_p}{dt} = \rho_{\rm part} v_{\rm rel} \pi R_p^2 F_g \tag{2.28}$$

where $\rho_{\rm part}$ is the density of the particle in the incoming stream. In case of constant F_g we have

$$\frac{1}{m_p}\frac{dm_p}{dt} \propto m_p^{-1/3} \tag{2.29}$$

Just because $m_p \propto R_p^3$ that once substituted in (2.29) gives a constant growth rate $\dot{R}_p = const$ and as consequence R_p will grow linearly with time. On the other hand in now $v_{\rm rel} = const$ then

$$\frac{1}{m_p}\frac{dm_p}{dt} \propto R_p \propto m_p^{1/3} \tag{2.30}$$

Which means a grow to $m_p \to \infty$ in a finite time and then a strong runaway. In order to obtain an estimate of the real grow rates we may assume as density ρ_p

$$\rho_p \approx \frac{\Sigma_p}{2H_p} \tag{2.31}$$

At this point to evaluate H_p we consider that in the disc $H_p = c_s/\Omega_k$ ad we just assume that $c_s \approx v_{\rm rel}$ so that finally

$$\rho_p \approx \frac{\Sigma_p \Omega_k}{2 v_{\rm rel}} \tag{2.32}$$

So making explicit the F_g factor we have

$$\frac{dm_p}{dt} = \frac{1}{2}\Sigma_p \Omega_k \pi R_p^2 \left[1 + \left(\frac{v_{esc}}{v_{\rm rel}}\right)^2\right] \tag{2.33}$$

This equation shows that the rate of growth is proportional to Σ_p while slows down for large distances (slower Ω_k and the relative velocity enters only in the focusing factor. With increasing population of planetesimals the surface density decreases and consequently the growth rate until it stops. The interactions between plantesimals become important and we ends up with protoplanets with dimension between the Moon and Mars. We will start from here to build our Earth.

2.4 The Construction of the Earth

The history of the Earth may begin already as a planetesimal. We do not need to emphasize that most of what we will write in this chapter is no more that a bunch of hypothesis corroborated by marginal evidence. Today the most likely history is summarized in **Fig.** 2.4 inspired to the work of Linda Elkins-Tanton and Kathryn Dodds and their coworkers. During the accretion of planetesimals up to hundred km may be partly differentiated and contains a highly viscous but liquid component that we call *magma ocean*. A magma ocean is a substantially part of the planet (≈10%) of molten material which may interest a portion of about 170 km depth on the Moon and 600 km on Earth. This magma ocean could be present already in the planetesimal produced both by the heat released by collisions but also by radiogenic heat produced by now extinct and short lived isotopes like ^{26}Al which has an half life of about 0.7 million year (Ma). Both planetesimals and planet embryos may be have differentiated and formed a core. An important role in these processes may be attributed to the collision between the Earth and an object of Mars dimensions that originated the Moon. We can fix some idea with energetic considerations. To assembly a planet a make it grow from radius m to $m + dm$ the energy required is

$$dE = -\frac{GM(r)}{r}dm \tag{2.34}$$

That once integrated gives the accretionary energy

$$E = -\frac{16}{15}\pi^2\rho^2 G R^5 \tag{2.35}$$

Which for the Earth amount to about 2×10^{32} J. This energy if released at once would heat the planet roughly by an amount $\Delta T \approx E/(c_p M)$ where c_p the heat capacity is roughly 800 J k^{-1} kg^{-1}. That would implies for the Earth an increase in temperature up to 40.000 K. However this heat is added on a timescale of 10 to hundreds million years and it delivered mainly from two sources. One is constituted by the impacts from planetesimals while the other source of energy is the formation of

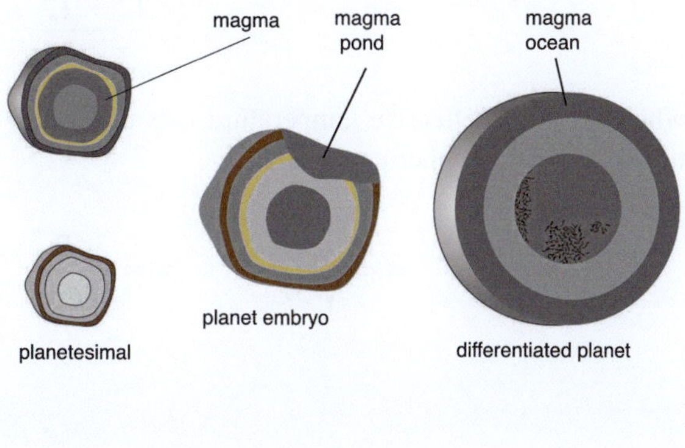

Fig. 2.4 The possible path to a differentiated Earth from accreting planetesimals to a planet embryo to a differentiated Earth

the core when a huge differentiation takes places with the iron component separating from the silicate and forms the nucleus of the Earth. This process releases potential energy. However part of the impact energy may be released as infrared radiation from the surface. At this point we just warn that the formation of the Earth could be a quite complicated matter and we will follow the rather simple approach that Yakata Abe and Toshihide Matsui described in a famous paper in 1985. They considered the growing Earth with surface temperature T_s and a surface layer of thickness d_s probably molten. This layer covered a subsurface layer d_h which received most of the heat from the impacts. They assume that the impacting planetesimals releases their heat to the subsurface layer of mass M_h with an efficiency h. If the temperature of the subsurface layer is T_h so that the energy balance for the subsurface layer is

$$h\frac{GM}{R}(M_h - M_s) = C_p(T_h - T_s)(M_h - M_s) \tag{2.36}$$

Here T_s is the temperature of surface layer M_s its mass, $C - p$ is the specific heat. The same equation for the surface layer becomes

$$(1 - h)\frac{GM}{R}M_s = 4\pi R^2 F_{\text{atm}} dt_s + C_p M_s (T_s - T_p) \tag{2.37}$$

Here now T_p is the temperature of the planetesimal before the impact and F_{atm} is the emitted heat (radiation) in the interval of time dt_s during which the planet grows from radius R to radius $R + d_s$. For simplicity assume $h = 0$ so that all the energy is released to the surface layer and $d_h = d_s$. We have now an interesting term in F_{atm} which implies the presence of some kind of atmosphere which produce a blanketing effect (greenhouse effect). This is provided by the water vapor released by the impacting planetesimals. This water vapor absorbs the infrared radiation and has an optical thickness given by

$$\tau = \frac{3}{2}\int_0^\infty k\rho_a dz \tag{2.38}$$

where k is the absorption coefficient. For a gray atmosphere (see appendix) the outgoing flux is

$$F_{\text{atm}} = 2\sigma\frac{T_s^4 - T_e^4}{\tau + 2} \tag{2.39}$$

where T_e is the effective temperature (see appendix). Now k can be expressed in terms of the atmospheric pressure

$$k = \sqrt{\frac{k_0 g}{3P_0}}, \qquad \tau = \frac{3kP}{2g} = \frac{3kM_a}{8\pi R^2} \tag{2.40}$$

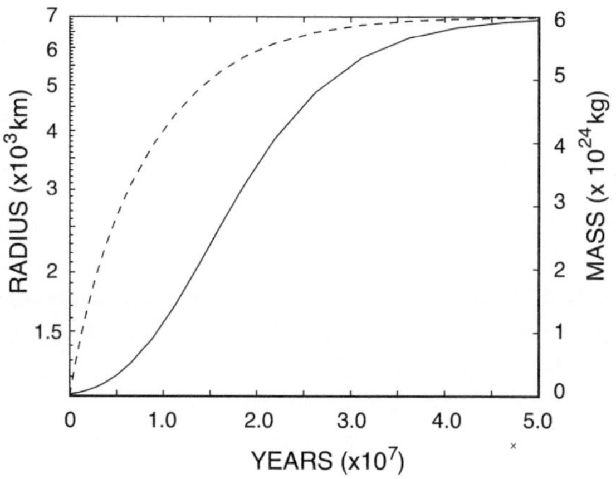

Fig. 2.5 Radius (dotted line) and mass (solid line) of the growing Earth. Notice that both quantities are slightly different from the real values

Where k_0 is the absorption coefficient at pressure P_0, P is the pressure at the surface and M_a the atmospheric mass. Although data can be updated we maintain the values adopted by Abe and Matsui with $k \approx 0.01\ \mathrm{m^2\ kg^{-1}}$ at a pressure $P_0 = 1.01 \times 10^5$ Pa. It can be shown that the atmospheric mass can be in the range of the present mass of the ocean. Consider now the growth of the planet base on the relation (2.24) we can write

$$\frac{dM}{dt} = \pi R^2(1+2\theta)\rho_p u \tag{2.41}$$

where θ is the so called Safronov factor

$$\theta = \frac{GM}{Ru^2} \tag{2.42}$$

ρ_p the density of planetesimal and u the impact velocity. Using (2.32) we get

$$\frac{dM}{dt} = (1+2\theta)\frac{2\Sigma_p\Omega_k}{\pi} \tag{2.43}$$

As the Earth grows it also consumes the accreting material so that the surface density decreases according to

$$\Sigma_p = \frac{M_0 - M}{\pi(a_0^2 - a_1^2)} + \Sigma_{p0} \tag{2.44}$$

where M_0 is the mass of the Earth, $\pi(a_0^2 - a_1^2)$ is the area of the collecting region and Σ_{p0} is the initial surface density. **Figure** 2.5 show the results of integration of (2.43) which corresponds to sweeping area of $\pi(a_0^2 - a_1^2)$ of about $4.08 \times 10^{22}\ \mathrm{m^2}$ and an initial mass of $2, 2 \times 10^{22}$ kg. The planet is completed in about 50 million years. Now we ask about the temperature. We have to distinguish between the surface and internal temperature. The surface temperature could be calculated according to (2.37) with $h = 0$ and then reads

$$c_p Ms \frac{dT_s}{dt} = \left[\left(\frac{2\theta + 1}{2\theta} \right) \frac{GM}{R} - c_p (T_s - T_p) \right] \frac{dM}{dt} - 4\pi R^2 (F_{\text{atm}} - F) \quad (2.45)$$

In this equation we remember M_s is the mass of surface layer, M is the mass of the planet, T_p is the planetesimal temperature, F_{atm} is the outgoing infrared flux while F is the flux from the Earth interior. The term in square brackets is the contribution to the surface temperature of the impacts while the last term contains a radiative term and the flux coming from the interior. The surface layer it is assumed has a thickness D_s given by an empirical formula as

$$D_s = b \left(\frac{V_i^2}{g} \right)^{\beta/3} \left(\frac{\rho}{M_{mprj}} \right)^{\frac{1}{3}\left(\frac{\beta}{3} - 1\right)} \quad (2.46)$$

where $\beta = 0.49$ and $b = 0.72$. M_{mprj} is the mass of the planetesimal and ρ its density. V_i is the impact velocity given by

$$V_i = \sqrt{\frac{GM}{R} \left(\frac{1 + \theta}{\theta} \right)} \quad (2.47)$$

Once D_s is calculated the mass of the surface layer is obtained as

$$M_s = 4\pi R^2 D_s \rho \quad (2.48)$$

And now (2.43), (2.45) can be solved as a system and the Matlab script is given in the appendix. The calculation is shown for both a planet without atmosphere where $F_{\text{atm}} = \sigma T_s^4$ and one where there is a primitive greenhouse effect where τ is not zero. The results are shown in **Fig.** 2.6. The enormous optical thickness implies a very large increase in temperature ($\approx$1500 K against $\approx$350 K. In the paper of 1985 Matsui and Abe made the hypothesis that water was released in the impact of planetesimals so that the increase of the mass of water vapor M_w is

$$\frac{dM_w}{dt} = \chi \frac{dM}{dt} \quad (2.49)$$

Then the increase in the mass of water vapor follows the same behavior of the mass of the planet and the pressure to be used in (2.40) is

$$P = \frac{M_w g}{4\pi R^2} = \frac{4}{9} \pi G \chi \rho R^2 \quad (2.50)$$

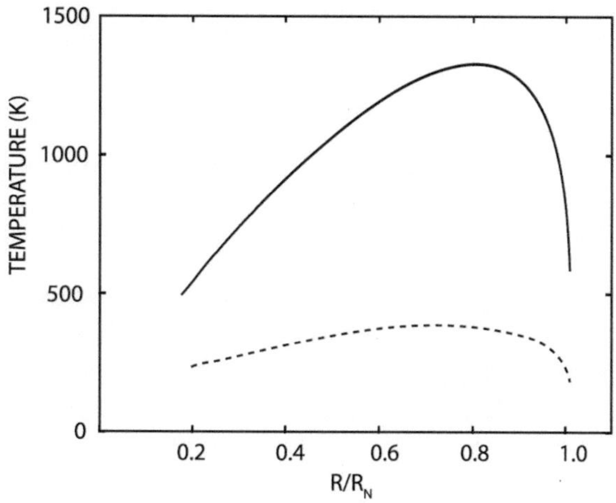

Fig. 2.6 The surface temperature of the Earth during its growth: with the greenhouse effect (solid line) and without the greenhouse (dotted line)

where we have substituted $g = GM/R^2$. So that as shown in **Fig.** 2.7 pressure increases with the square of the radius of the planet. With $\chi = 2 \times 10^{-4}$ we reach a pressure of roughly 18 MPa. These high temperature and pressures may generate what is known as *magma ocean*. As a matter of fact shown already by William Kaula in his seminal paper of 1979 the surface can exceeds the *solidus temperature* that is begin to melt. The expression used for the solidus is with pressure P in Pascal.

$$T_{\mathrm{sol}} = 1500 + 5.47 \times 10^{-8} P - 2.17 \times 10^{-19} P^2 \tag{2.51}$$

And the liquidus temperature usually is 200 K higher. In their paper of 1986 Abe and Matsui showed that the mass of the atmosphere reached the value of 1.9×10^{21} kg (the present ocean mass is $\approx 1.4 \times 10^{21}$ kg) and the pressure of the order of 10^7 Pa. At this point the surface would have reached the solidus temperature and began to melt and as required by the physics the temperature would stay constant. It would be the start of the magma ocean, however before discussing that we would like to mention what happened of the internal thermal structure of the planet.

2.5 The Thermal Structure of the Growing Earth

The reader may be convinced that things on the primitive Earth went as described. But it is clear that we have made a lot of simplification. The planet is heated on the surface by the energy dissipated of the bombardments, But now some heat is being provided by the decay of the radioactive elements and a huge source of energy is the differentiation which generates the iron core. So we simplify the problem just reducing it to a spherical planet heated by radioactivity and planetesimal bombard-

Table 2.1 Energy output for long lived radioactive isotopes

Isotope	C_i $10^{-8} gg^{-1}$	A_i $\mu W kg^{1}$	λ_i $10^{-10} y^{-1}$	$\exp\lambda_i t^*$
^{238}U	1.99	95	1.55	2.04
^{235}U	0.01	562	9.84	92.4
^{232}Th	7.0	26	0.49	1.25
^{40}K	4.67	30	5,54	12.8

ment and we assume that the dominant mechanism of heat transport is diffusion. So we write the diffusion equation in spherical coordinates

$$\frac{\partial T}{\partial t} = \frac{1}{r^2}\frac{\partial}{\partial r}\left[r^2 k \left(\frac{\partial T}{\partial r}\right)\right] + \frac{E}{\rho c_p} \tag{2.52}$$

In this equation k is the heat diffusivity while E is the heat generation by the decay of radioactive elements (E_R) and the energy released by the impacts (E_I). The latter implies quite involved hypothesis and calculation and the reader may refer to the 1983 paper by Angioletta Coradini while the radioactive term is much more simple because is a function of the concentration of the four isotopes $^{238}U, ^{235}U, ^{232}Th, ^{40}K$. So the heat generation by radioactive decay is characterized by the lifetime λ_i of the different isotopes, their concentration C_i and the rate of heat generation A_i. These quantities are listed in **Table** 2.1. Notice that the abundance of the various elements at the time of the origin of the planet was larger by a factor $exp(\lambda_i t^*)$ where t^* is the age of the Earth ($\approx 4, 6$ billion years). The heat generation can be written as

$$E_R = \sum C_i A_i exp(-\lambda_i t) \tag{2.53}$$

The summation is done on the long lived isotopes listed below and where the concentration C^*, t^* years ago is given by $C_i^* = C_i exp(\lambda t^*)$.

As the Earth grows the inner layers are compressed and this raise the temperature according to the adiabatic compression

$$\Delta T_{ad}/r = \frac{4\pi G \epsilon R \rho}{3 c_p} T(r) \Delta R \tag{2.54}$$

where ϵ is the thermal expansion coefficient and ΔR is the increment in radius. Av way to obtain (2.54) i shown in the appendix. Again some hypothesis has to be made concerning the diffusivity but in any case (2.53) can be solved. This approach may look quite complicated but it is actually very much simplified because it neglects the energetic of the core formation and the heat transport from the interior that we mention in the next paragraph.

2.6 Magma Ocean and the Primordial Atmosphere

We already mentioned that the high temperature of the Earth surface reached during its formation may result in the melting of a deep layer producing what is known a magma ocean. There is a simple calculation reported by Chusiro Hayashi, that we repeat here with some minor change. He consider that the energy released due to the impact is given by

$$E_i = \frac{GM}{R}\dot{M} \tag{2.55}$$

where M is the mass of the Earth and $\dot{M}$ its increment in time. Considering that the growth time for the planet is τ_G we have roughly $\dot{M} = M/\tau_G$. And so the impact energy will be only a function of the Earth radius

$$\frac{E_i}{4\pi R^2} = G\frac{M^2}{4\pi R^3 \tau_G} = \frac{4\pi}{9}G\frac{R^3\rho^2}{\tau_G}. \tag{2.56}$$

This input energy must be balanced by the infrared flux (2.39) with the optical thickness evaluated according (2.40) so that we need to specify the ratio χ between the atmospheric mass and the mass of the planet. Notice that now pressure is a function of $g = GM/R^2$ so that change during the formation of the planet. The temperature of the surface is then

$$T_s^4 = \frac{2\pi G}{9\sigma}\frac{R^3\rho^2}{\tau_G}(\tau + 2) + T_e^4 \tag{2.57}$$

Figure 2.6 shows that the very high temperature can be reached at the surface while the pressure of the atmosphere could be (see **Fig.** 2.7 in the range of several Mpa. These conditions are such and the upper layer (of hundred km thickness) may melt resulting in what is known as magma ocean. The temperature at the surface may be maintained not only by the greenhouse effect but also by the blast of planetesimal. Roughly we can assume that the energy dissipated in a single collision goes into heating of the ocean according to the relation

$$\frac{1}{2}mv^2 = 4\pi R^2 \epsilon D\rho C_p \Delta T \tag{2.58}$$

where m is the mass of the planetesimal. v its impact velocity, ϵ the efficiency, D the depth of the ocean, ρ and C_p the density and specific heat and ΔT the increase in temperature. The numerical value may refer to 95% of the Earth mass and a radius of $\approx$ 6900 km an efficiency of 0.5, and $C_p \approx 1250$ J kg^{-1} K^{-1}. The calculations were carried out with a distribution of masses like $n(m) \approx m^{\cdot\alpha}$ in a short paper by Stevenson and he showed that with a constant pressure of about 46 bar the surface could be maintained at about 1350 K. The most probable composition of the primordial atmosphere is mainly water vapor because in meteorites we have as much as 20%

Fig. 2.7 Atmospheric pressure of growing Earth

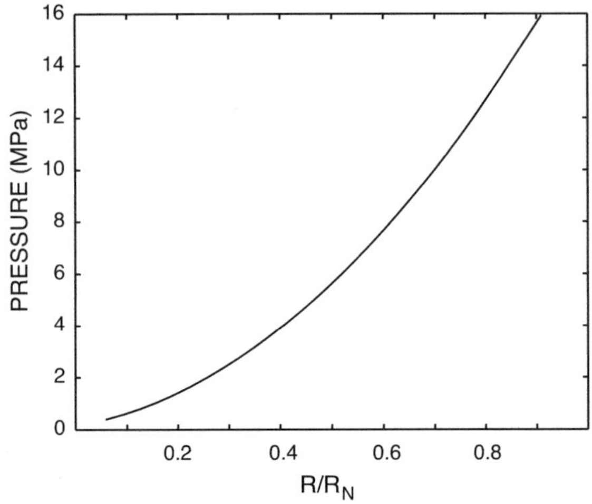

of water. Because of the large solubility in a 100 bar atmosphere about 1% of water could dissolve. If the pressure of the H_2O atmosphere decrease because of the dissolution in the silicate melt the blanketing effect of the atmosphere decreases and the surface cool lowering the content of water. When this content becomes smaller than the source material (i.e. meteorites) the abundance of H_2O in the proto atmosphere increases, because supply by impact degassing surpasses the loss by dissolution. This negative feedback mechanism it is just right to maintain the magma ocean and presumably the water vapor content at the end of the accretion period reaches a pressure of about 300 bar that is a mass of roughly $\approx 10^{21}$ kg equivalent to the present mass of the oceans. The presence of the magma oceans give raises to the issue of the heat transported through the thickness of the magma ocean to the surface. We show in the appendix that such a flux can be expressed through the so called *Nusselt number* defined as the ratio of the convective flux to the conductive flux. If q is the total flux at the top of the layer we have

$$Nu = \frac{qD}{K\Delta T} = \frac{q}{q_c} \tag{2.59}$$

where q_c is the conductive flux, D the thickness of the layer. K the thermal conductivity and ΔT the temperature difference. It can be shown that the Nusselt number can be related to the Rayleigh number R_a as

$$Nu = a\left(\frac{R_a}{R_{ac}}\right)^p \tag{2.60}$$

With $a \approx 1$ and $p \approx 1/3$. and the Rayleigh number

$$R_a = \frac{g\alpha D^3 \Delta T}{k\nu} \tag{2.61}$$

where g is the gravity acceleration α the thermal expansion coefficient, D the depth, k the thermal diffusivity $k = K/\rho C_p$ and ν the kinematic viscosity $\nu = \eta/\rho$ with η the viscosity of the fluid. Geoffrey Davies in a nice paper in 1985 found a very ingenious way to express the convective flux q: For simplicity he assumes $T = \Delta T$ and then makes the ratio between q ans some reference flux q_0 must be equal to the ratio of the Nusselt numbers and then of the Rayleigh numbers

$$\frac{q}{q_0}\frac{T_0}{T} = \left(\frac{T}{T_0}\right)^p \left(\frac{\nu_0}{\nu}\right)^p \rightarrow \frac{q}{q_0} = \left(\frac{T}{T_0}\right)^{p*1} \left(\frac{\nu}{\nu_0}\right)^{-p} \tag{2.62}$$

where we consider only quantities which depends on T. In particular we consider that the kinematic viscosity depends on T

$$\frac{\nu}{\nu_0} = \left(\frac{T}{T_0}\right)^{-n}. \tag{2.63}$$

With n a quite large number $\approx$30 so that the flux

$$\frac{q}{q_0} = \left(\frac{T}{T_0}\right)^m \qquad m = 1 + (n+1)p \approx 11 \tag{2.64}$$

This very simple relation can be applied to the study of the planet cooling and that of the magma ocean. We assume the planet has a mass M, a specific heat c_p so that its heat content will be $Q = Mc_pT'$ where T' is some temperature that will make non dimensional with some reference temperature T_0'. So the heat balance equation will be

$$\frac{\partial Q}{\partial t'} = c\frac{\partial T'}{\partial t'} = q' - A \tag{2.65}$$

where A is the additional radioactive heat. All the quantities appearing in (2.65) can be normalized as follows.

$$T = \frac{T'}{T_0'}, \qquad q = \frac{q'}{q_0'}, \qquad h = \frac{A}{q_0'}, \qquad t = \frac{t'}{\tau} \tag{2.66}$$

with τ

$$\tau = c\frac{T_0'}{q_0'} \tag{2.67}$$

So the non dimensional version of (2.65) becomes

$$\frac{\partial T}{\partial t} = h - q = h - T^m \tag{2.68}$$

We can study the solutions for $m = 1$ and $h = 0$ (no heat sources). In thid case we have an analytical solution like

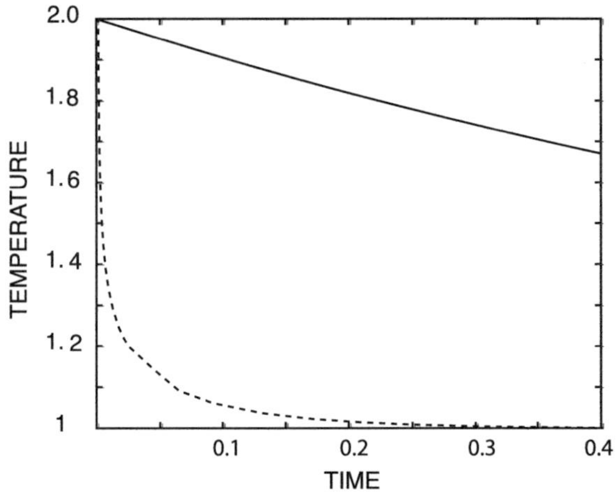

Fig. 2.8 Cooling of a planet with convection (dotted line) and without convection (solid line)

$$T = e^{-t} = e^{-t'/\tau} \tag{2.69}$$

With the idea that $T = 1$ for $T = 0$. So the decay constant τ is of the order of 6×10^9 years when $M = 4 \times 10^{24}$, $c_p = 1000$ J kg^{-1} K^{-1}, $T_0 = 2000$ K and $q \approx 4 \times 10^{13} w$. That is without convection the ocean or the mantle would cool in a time longer that the edge of the Earth. When we introduce convection the solution becomes

$$\frac{1}{T^{m-1}} = 1 + (m-1)t \tag{2.70}$$

This solution implies that for $t = -1/(m-1)$ that is for $t' = -\tau/(m-1)$ the temperature would go to a very large value. This time is about 0.6 billion years. This thermal catastrophe requires to consider a heat source h when the convection is working. The results of this exercise are shown in **Fig.** 2.8 just to see that cooling is much faster when convection works. One of the first conclusions that came to mind is that the surface of the growing planet gets hot enough to melt the silicate and the water vapor release makes up an atmosphere dense enough that when condense can result in an ocean of water. The details of this process will be examined in a while and we notice that the present mass of the ocean 1.3×10^{21} kg would correspond to an atmosphere with a ground pressure of roughly 25 MPa. This amount of water vapor will be a rather important infrared blanket that could keep the surface temperature above 1400 K because of an exasperate runaway greenhouse that we explain in the next paragraph.

2.7 The Runaway Greenhouse

The concept of runaway greenhouse was invented in 1966 paper by Makoto Komabayasi and elaborated three years later by Andrew Ingersoll. The basic idea is that if the solar flux increases the surface temperature increases and more water vapor enters the atmosphere which increases the optical thickness.

This positive feedback will continue up to the point that nowhere in the atmosphere there are conditions (pressure and temperature) for the vapor to condense: in this way the whole oceans could evaporate. This probably happened to a planet like Venus, Both Komabayasi and Ingersoll played with the solar flux but something similar could happens by increasing the optical thickness of the atmosphere. The original paper started with the grey atmosphere solution for the surface temperature (see appendix)

$$\sigma T^4 = \frac{F_0}{2}\left(1 + \tfrac{3}{2}\tau\right), \quad \rightarrow \tau = \tfrac{4}{3}\left(\frac{\sigma T^4}{F_0} - 1\right) = \frac{4\sigma}{3F_0}(T^4 - T_0^4) \tag{2.71}$$

here we have used that in the absence of the atmosphere $\sigma T_0^4 = F_0$. Then remembering that $\tau = ku = k(ug)/g$ with u expressed in kgm^{-2} then $ug = e$ is the vapor pressure so we have

$$e = \frac{4\sigma g}{3F_0 k}(T^4 - T_0^4) \tag{2.72}$$

This can be compared with the saturation pressure

$$e_s = Ae^{\beta T} \tag{2.73}$$

With $A = 611\ Pa$ and $\beta = 0.067$ and T in degrees centigrade. This is done in **Fig.** 2.9. We see that of the two intersection points T_1 and T_2 only the first is stable because with any increase in temperature the pressure e remains higher then e_s. This

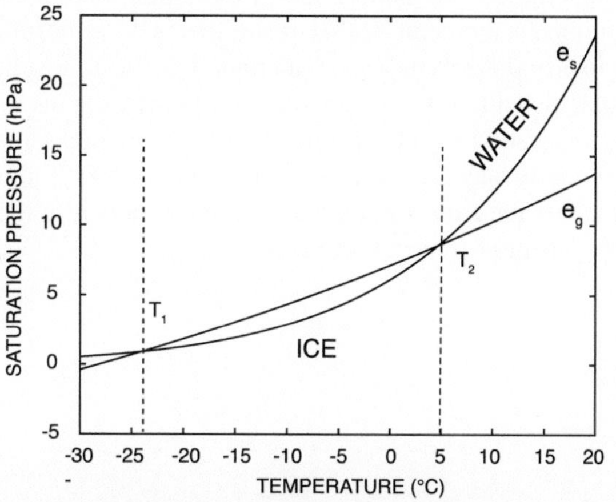

Fig. 2.9 Radius (dotted line) and mass (solid line) of the growing Earth. Notice that both quantities are slightly different from the real values

up to the temperature T_2 when for any increase in temperature $e > e_s$: is the runaway greenhouse. The treatment given by Ingersoll is little more elaborate. He starts by defining the optical thickness

$$\tau = \int \rho_v K dz = \int \chi \rho K dz = \int \frac{e_s}{p} \frac{m_v}{m} K \rho dz = -\int K \frac{e_s}{p} \frac{m_v}{m} \frac{dp}{g} \tag{2.74}$$

where $\chi = \rho_v/\rho$ is the mass mixing ratio defined as the ratio between the density of water vapor ρ_v and that of air ρ. In (2.74) we have used the hydrostatic equilibrium so that $\rho dz = -dp/g$. At this point the assumption is that the relative humidity $e_s/p = 100\%$ all over so that finally the optical thickness is

$$\tau = \frac{K m_v e_s(T)}{mg} \tag{2.75}$$

Because the saturation pressure e_s is only function of temperature so becomes also the optical thickness. As we show in the appendix at the radiative equilibrium in a grey atmosphere the temperature as a function of optical thickness is given by (2.71)

$$\sigma T^4 = \frac{F_0}{2} \left(\frac{3}{2} \tau + 1 \right)$$

where F_0 is the incoming solar flux. Now we can substitute (2.75) in (2.76) and obtain an implicit relation for the temperature T in terms of the flux F_0

$$\frac{F_0}{2\sigma} = T^4 \left(\frac{3}{2} \tau + 1 \right)^{-1} = T^4 \left[1 + \frac{e_s(T)}{P_0} \right]^{-1}, \quad P_0 = \frac{2mg}{3Km_v} \tag{2.76}$$

To explicate the dependence of τ from T we can substitute (2.73) in (2.75) and produce a Figure like **Fig.** 2.10 where we superimpose (2.71)–(2.75). This is done for different values of the solar flux (185, 285, 385, 485 w m^{-2}). we see that beyond 385 w m^{-2} there is no intersection of the saturation optical thickness with the radiative equilibrium temperature that means that in the atmosphere there are not condensation regions and the water vapor can remain in the atmosphere as a a gas. There are a few problems with this simple model of runaway the most important being that the gray approximation is not valid for the very large pressures we envisaged. Also another hypothesis may proves to be wrong and that is the radiative equilibrium solution and we rather must use a radiative convective model for the lower atmosphere. However the concept of runaway survives.

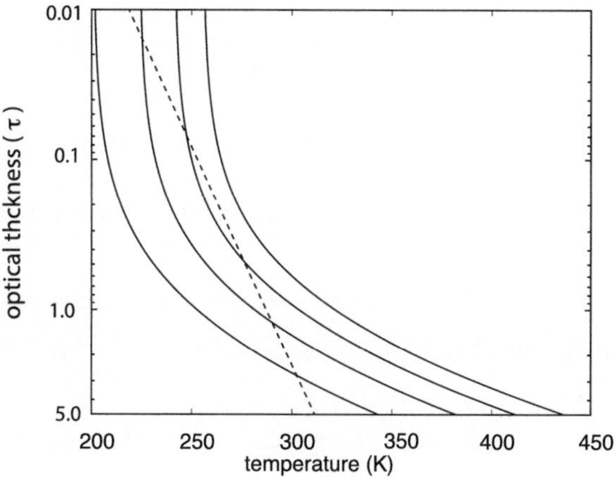

Fig. 2.10 The runaway greenhouse. The temperature as a function of optical thickness for solar flux of 185, 285, 385 ans 485 $\mathrm{w\,m^{-2}}$ superimposed to the saturation pressure as function of optical thickness (dotted line). It is evident that the highest value of solar flux leads to runaway greenhouse

Appendix

The Gravitational Instability

The first paper about gravitational instability was published in *The Astrophysical Journal* in 1954 by Subrahmanyan Chandrasekhar. Actually the idea was to generalize what was called Jeans (from his discoverer James Jeans) instability by including the presence of rotation and magnetic field. The originality of the three pages paper was that it made a perturbation analysis of a gas cloud when it was disturbed by a density wave. The question is under what conditions such a wave could be amplified. When referred to the instability in the disc we just write the equations of continuity, momentum and Poisson

$$\begin{aligned}\frac{\partial \Sigma}{\partial t} + \nabla \cdot (\Sigma \boldsymbol{v}) &= 0 \\ \frac{\partial \boldsymbol{v}}{\partial t} + (\boldsymbol{v} \cdot \nabla)\boldsymbol{v} &= -\frac{\nabla p}{\Sigma} - \nabla \Phi \\ \nabla^2 \Phi &= 4\pi G \rho\end{aligned} \tag{2.77}$$

We assume that the equilibrium solution to these equations is given by

$$\begin{aligned}\rho_0 &= \Sigma \delta(z) \\ \boldsymbol{v_0} &= (0, r\Omega, 0) \\ \Omega^2 r &= -\frac{d\Phi_0}{dr}\end{aligned} \tag{2.78}$$

Now we can make a linear stability analysis with a small perturbation on these values

$$
\begin{aligned}
\Sigma &= \Sigma_0 + \Sigma'(r,t) \\
p &= p_o + p'(r,t) \\
\Phi &= \Phi_0 + \Phi'(R,T) \\
\boldsymbol{v} &= \boldsymbol{v_0} + [v_r(r,t), \delta v_\phi(r,t), 0]
\end{aligned}
\tag{2.79}
$$

Assume now that all the perturbation have the form

$$\Sigma'(r,t) \propto exp(kr - \omega t) \tag{2.80}$$

So that the first of (2.77) becomes

$$-i\omega\Sigma' + v_r\Sigma_0\left(\frac{1}{r} + ik\right) \approx -\omega\Sigma' + kv_r\Sigma_0 = 0 \tag{2.81}$$

where the approximation is valid in the limit $kr \gg 1$. To apply the perturbation to the momentum equation the only problem is to express then in cylindrical coordinates. So at the end we have

$$
\begin{aligned}
-i\omega v_r - 2\Omega\delta v_\phi &= -\frac{1}{\Sigma_0}\frac{dp'}{dr} - \frac{d\Phi'}{dr} \\
-i\omega\delta v_\phi + v_r\left[\Omega + \frac{d}{dr}(r\Omega)\right] &= 0
\end{aligned}
\tag{2.82}
$$

Notice that the first term on the right of the first of (2.82) can be expressed in terms of the sound velocity

$$\frac{1}{\Sigma_0}\frac{dp}{dr} = \frac{1}{\Sigma_0}\frac{dp}{d\Sigma}\frac{d\Sigma}{dr} = \frac{1}{\Sigma_0}c_s^2 ik\Sigma' \tag{2.83}$$

To work on the Laplacian we express it in cylindrical coordinates

$$\frac{1}{r}\frac{\partial}{\partial r}\left(r\frac{\partial\Phi}{\partial r}\right) + \frac{1}{r^2}\frac{\partial^2\Phi}{\partial\phi^2} + \frac{\partial^2\Phi}{\partial z^2} = 0 \tag{2.84}$$

First of all the perturbed Poisson equation gives

$$\nabla^2\Phi' = 4\pi\Sigma'\delta(z) \tag{2.85}$$

Assuming for Φ' a dependence like (2.80) we write

$$\frac{d^2\Phi'}{dz^2} - k^2\Phi' = 4\pi\Sigma'\delta(z) \tag{2.86}$$

The solution to this equation which remain finite for large z has the form

$$\Phi'(z) = C \exp -|kz| \tag{2.87}$$

Integrating on z around the plane $C = -2\pi G\Sigma'/|k|$ so finally we get

$$\Phi' = -\frac{2\pi G\Sigma'}{|k|} \tag{2.88}$$

Taking the derivative with respect to r we have

$$\frac{d\Phi'}{dr} = -\frac{2\pi ikG\Sigma'}{|k|} \tag{2.89}$$

Substituting this result and (2.83) in the first of (2.82) we have

$$-i\omega v_r - 2\Omega\delta v_\phi = -\frac{c_s^2 ik\Sigma'}{\Sigma_0|k|} + \frac{2\pi ikG\Sigma'}{|k|} \tag{2.90}$$

We can now eliminate v_r and v_ϕ using (2.81) and the second of (2.82) to get the *dispersion relation* which express ω as a function of k

$$\omega^2 = \kappa^2 + c_s^2k^2 - 2\pi G\Sigma_0|k| \tag{2.91}$$

where κ is known as *epicyclic frequency*

$$\kappa^2 = 4\Omega^2 + 2r\Omega\frac{d\Omega}{dr} \tag{2.92}$$

In case of keplerian motion $\Omega = GM/r^2$ so that

$$\Omega^2 = \frac{GM}{r^3} \rightarrow \frac{d\Omega}{dr} = -\frac{3}{2}\frac{\Omega}{r}$$

so that $\kappa^2 = \Omega^2$. The instability occurs for all those values of k that gives negative values for ω^2 that means imaginary values for ω. To find the most unstable wavelength we perform $d\omega^2/dk = 0$ to find

$$k_{\min} = \frac{\pi G\Sigma_0}{c_s^2} \tag{2.93}$$

So that for the disk to be marginally stable the requirement is $\omega^2(k_{\min} = 0$ and for a Keplerian disc

$$\omega^2 = \frac{\Omega^2c_s^2 - \pi^2G^2\Sigma_0^2}{c_s^2} < 1 \rightarrow Q \equiv \frac{c_s\Omega}{\pi G\Sigma_0} < 1 \tag{2.94}$$

This long procedure may be useful to find the typical dimension of planetesimals originating from gravitational instability. We observe that the most unstable wavelength is

$$\lambda \approx \frac{2\pi}{k_{\min}} \approx \frac{2c_s^2}{G\Sigma_0} \tag{2.95}$$

substituting c_s from (2.94) we obtain an order of magnitude for the mass of the planetesimal

$$m_p \approx \pi\lambda^2\Sigma_s \approx \frac{4\pi c_s^4}{G^2\Sigma_s^2} \approx 4\pi^5 G^2 Q^4 \frac{\Sigma_s^3}{\Omega^4} \tag{2.96}$$

Using $Q = 1$, $\Sigma_s \approx 10^2$ kg m^{-2} we obtain

$$m_p \approx 3 \times 10^{15} \text{ kg} \tag{2.97}$$

For a distance of 1 A.U. (i.e. $\Omega = 2\pi/3.15 \times 10^7$ s^{-1}). This mass corresponds to a sphere with radius of 3.5 km and $\lambda \approx 3$ km.

The Growth of Particles by Sedimentation

We like to solve (2.13) coupled with (2.14)

$$\frac{dm}{dt} = \frac{3}{4}\frac{\Omega^2 f}{\bar{v}} zm$$
$$\frac{dz}{dt} = \frac{\rho_d}{\rho}\frac{\Omega^2 f}{\bar{v}} za$$

The Matlab script which solves the two equation is given below.

```
[t,x]=ode45('growth1',[0 3.15e10],[15e11 1.25e-14]);
rom=3;
figure(1)
semilogy(t/3.15e7,x(:,1)/3e11),axis([0 1100 0.01 10])
figure(2)
rom=3;
wk1=(3*x(:,2)/(4*pi*rom)).^(1/3);
%semilogy(t/31.5,x(:,2))
semilogy(t/3.15e7,1.e4*wk1),axis([0 1100 0.01 1.e4])
function xdot=growth1(t,x);
OM=2*pi/3.15e7;
f=0.01;
h=3e11;
```

```
xdot=zeros(2,1)
vth=3.e5;
rod=3.;
ro=6.e-10;
xdot=zeros(2,1);
A=rom*OM^2/(ro*vth);
B=0.75*f*(OM^2/vth);
a=(3*x(2)/(4*pi*rod))^(1/3);
xdot(1)=-A*a*x(1);
xdot(2)=B*x(1)*x(2);
```

The script is self explanatory. In any case integration is carried out for 1000 year (3.15×10^{10} s) with initial values of 5 h with $h = 3 \times 10^{11}$ cm and a mass of $1.25 \times 10^{-}14$ g which correspond to a particle of 0.1 μm radius. The terminal velocity vth is 3×10^5 cm s^{-1}. The results of such integration are given in **Fig.** 2.1 and they are different from those of Fig. 4.1 of the Armitage text although ours looks much more coherent. Aa a matter of fact they show a slow descent of the particles in the initial stage when the particle is still quite light.

The Grey Atmosphere

We consider a plane parallel atmosphere where we define an intensity of radiation $I(\tau, \mu)$ as the power per unit surface and unit solid angle in the direction defined by $\mu = cos\theta$ (where θ is the zenith angle) and τ the optical thickness increasing from $z = \infty$ and defined as

$$\tau = \int_z^\infty k\rho_v dz \tag{2.98}$$

With k absorption coefficient for the gas of density ρ_v. The transfer of radiation is defined by the equations

$$\begin{aligned} \mu\frac{dI(\tau, \mu)}{d\tau} &= I(\tau, \mu) - B(T) \\ -\mu\frac{dI(\tau, -\mu)}{d\tau} &= I(\tau, \mu) - B(T) \end{aligned} \tag{2.99}$$

where $\sigma T^4 = \pi B(T)$ with σ Stefan Boltzmann constant at this point we define an upward F^+ and downward F^- flux as

$$F^+ = 2\pi \int_0^1 \mu I(\tau, \mu)d\mu, \qquad F^- = 2\pi \int_0^1 \mu I(\tau, -\mu)d\mu \tag{2.100}$$

For isotropic radiation (the same in all direction) we get

$$F^+ = \pi I^+, \qquad F^- = \pi I^- \tag{2.101}$$

With I^+, I^- the isotropic intensity in the upper and lower hemisphere. Carrying out the averaging on (2.99) we get

$$\begin{aligned} \frac{2}{3}\frac{dF^+}{d\tau} &= F^+ - \pi B \\ -\frac{2}{3}\frac{dF^-}{d\tau} &= F^- - \pi B \end{aligned} \tag{2.102}$$

Sometime people re define the thickness as $\tau^* = \frac{3}{2}\tau$ so that (2.102) simplify as

$$\frac{dF^+}{d\tau} = F^+ - \pi B, \qquad -\frac{dF^-}{d\tau} = F^- - \pi B \tag{2.103}$$

These equations can be easily solved by summing and subtracting with the boundary condition that $F^-(0) = \sigma T_e^4$. We also observe that at *radiative equilibrium* the net flux $F_{\text{atm}} = F^+ - F^- = const$ (so that $dF_{\text{atm}}/d\tau = 0$). We arrive at the results

$$F^+(\tau) = \frac{F_{\text{atm}}}{2}(\tau + 2) + F^-(0), \qquad F^-(\tau) = \frac{F_{\text{atm}}}{2}\tau + F^-(0) \tag{2.104}$$

where

$$F_{\text{atm}} = 2\sigma \frac{T_s^4 - T_e^4}{\tau_s + 2} \tag{2.105}$$

With T_s and τ_s the temperature and thickness at the ground. We also have

$$\pi B(\tau) = \sigma T(\tau)^4 = \frac{F_{\text{atm}}}{2}(\tau + 1) + F^-(0) \tag{2.106}$$

Which gives the temperature profile as a function of τ. At the surface the emitted flux coincide with upward flux (2.104)

$$\sigma T_s^4 = \frac{F_{\text{atm}}}{2}(\tau_s + 2) + F^-(0) \tag{2.107}$$

Finally the effective temperature T_e is given by $\sigma T_e^4 = S_0$ where S_0 is the net incoming solar radiation.

The Growth of the Earth

This paragraph is to present a Matlab script which solves Eq. (2.43). The radius is obtained assuming a constant density.

```
[t,x]=ode45('growth1',[0 3.15e10],[15e11 1.25e-14]);
rom=3;
figure(1)
semilogy(t/3.15e7,x(:,1)/3e11),axis([0 1100 0.01 10])
figure(2)
rom=3;
wk1=(3*x(:,2)/(4*pi*rom)).^(1/3);
%semilogy(t/31.5,x(:,2))
semilogy(t/3.15e7,1.e4*wk1),axis([0 1100 0.01 1.e4])
function ydot=abemod(t,y)
ro=4200;
teta=4;
m0=6e24;
Pk=3.15e7;
area=4.08e22;
R=(3*y/(4*pi*ro))^(1/3);
sigp=(m0-y)/area;
ydot=3.15e7*pi*R^2*9*4*sigp/Pk;
```

A similar script follows which deals with the heating of the planet due to the bombardment of planetesimals.

```
[t,x]=ode45('planetheat',[0 5e7],[2.3e22,300]);
r=(0.75*x(:,1)/(pi*4200)).^(1/3);
rn=r/6.9e6;
G=6.67e-11;
rho=4200;
f=2.e-4;
press=(4/9)*pi*G*f*rho^2*r.^2;
figure (1)
plot(rn,x(:,2))
figure(2)
plot(rn,press)

function xdot=planetheat(t,x);
xdot=zeros(2,1);
teta=4;
sib=5.67e-8;
area=4.08e22;
Mf=6e24;
sp=(Mf-x(1))/area;
Pk=3.15e7;
ro=4200;
xdot(1)=3.15e7*pi*(0.75*x(1)/(pi*ro))^(2/3)*(1+2*teta)*4*sp/Pk;
```

```
R=(0.75*x(1)/(pi*ro))^(1/3);
b=0.72;
beta=0.49;
G=6.67e-11;
ro=4200;
g=9.8;
cp=1000;
Mprg=1.e15;
V1=(2*G*x(1)/R+G*x(1)/(teta*R))^(1/2);
Tp=250;
help=(beta/3-1)/3;
Dcons=b*(V1^2/g)^(beta/3)*(ro/Mprg)^help;
Ms=4*pi*R^2*Dcons*ro;
ko=5.7e-4;
rr=R/6.9e6;
f=2e-4;
pw=(4/9)*pi*G*f*ro^2*R.^2;
gr=G*x(1)/R^2;
k=sqrt(ko*g/(3*1e5));
tau=1.5*k*pw/gr;
Fatm=2*sib*(x(2)^4-150^4)/(tau+2);
werk1=((9/8)*G*x(1)/R-cp*(x(2)-Tp))*xdot(1)-3.15e7*4*pi*R^2*0.75*Fatm;
xdot(2)=werk1/(cp*Ms);
```

In the program also the pressure is calculated by assuming that its mass is proportional to the accreting mass with a ratio $f = 2 \times x10^{-4}$.

The Adiabatic Heating

From the classical text of William Kaula we know that for an adiabatic compression in a solid we have

$$\left(\frac{dT}{dP}\right)_s = \frac{\alpha T}{\rho C_v} \tag{2.108}$$

where α is the thermal expansion coefficient, C_v then specific heat at constant volume and ρ the density. The pressure at radius r within a planet of radius R can be calculated by considering the pressure of a thin layer over a sphere o radius r

$$\frac{4\pi\rho r^2 drg}{4\pi r^2} = \frac{4}{3}\pi\rho^2 Grdr \tag{2.109}$$

where we have substituted $g = Gm/r^2$. Integrating between limits R (the radius of the planet and r we have

$$P(r) = \frac{2\pi}{3}\rho^2 G(R^2 - r^2) + p_a \tag{2.110}$$

from which we have

$$dP(r) = \frac{4\pi\rho^2 GR}{3} dR \tag{2.111}$$

Using now (2.108) we obtain

$$\Delta T(r) = \frac{4\pi\alpha T(r)\rho GR}{3C_v} \Delta R \tag{2.112}$$

We notice that this formula coincide with the relation (11) in the paper by Abe and Matsui (1995) while we could not reconstruct the relation (59) in Kaula (1969) that seems to be incorrect even dimensionally. Equation (2.112) can be interpreted as the additional potential energy $4\pi R^2 \rho \Delta RG/R$ which heats the layer.

The Heat Transport in the Magma Ocean

It is necessary to give a short introduction to the problem of heat transport in fluid. That is based on two different mechanisms, one being the movements of the fluid (that is its velocities) and the other being the diffusion also called conduction. The equation describing the first mechanism id expressed by the total derivative

$$\frac{dT}{dt} = \frac{\partial T}{\partial t} + u\frac{\partial T}{\partial x} + v\frac{\partial T}{\partial y} + w\frac{\partial T}{\partial z} = \frac{\partial T}{\partial t} + \boldsymbol{v} \cdot \nabla T \tag{2.113}$$

The term $\boldsymbol{v} \cdot \nabla T$ also called *advective* term and in a sense represents the heat transport from neighboring space. O the other had conduction or diffusion can be represented by the diffusion equation

$$\frac{\partial T}{\partial t} = \kappa \frac{\partial^2 T}{\partial x^2} \rightarrow \kappa \nabla^2 T \tag{2.114}$$

where κ is the diffusion coefficient with dimension $m^2\, s^{-1}$. A quite intuitive approach to convection in a fluid is given in the beautiful book by Davies. He consider convection in the Earth mantle which is responsible of the plate tectonics as illustrated in **Fig.** 2.11. The process can be simplified as shown in the insert with a plate rising to the left due to the buoyancy and one descending on the right due to the cooling. We evaluate the buoyancy of the ascending portion due to decrease in density $\Delta\rho$ which is given by

$$B = gDd\rho\alpha\Delta T \approx -gDd\rho\alpha\frac{T}{2} \tag{2.115}$$

where D is the length of the plate, d its thickness, ρ the density ad $\Delta T = -T/2$ the temperature difference. Notice that the depth dimension is missing in (2.113) and it is assumed to be unitary. So that (2.113) gives the force for unit length because is obtained as product of the acceleration ($g\alpha\Delta T$) for the mass for unit volume (ρ) for

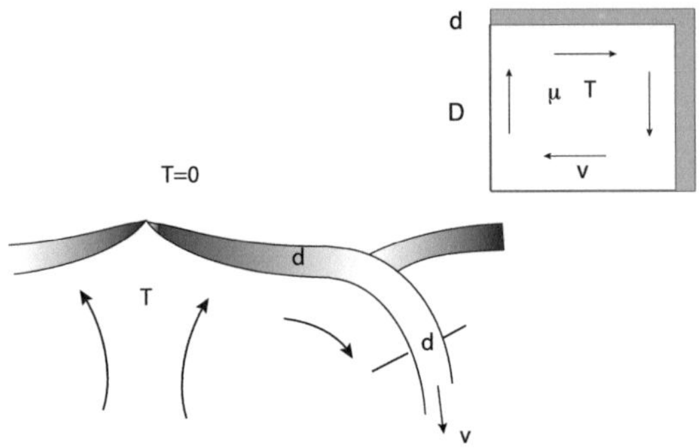

Fig. 2.11 The schematic convection

the volume (Dd). In the same way for the descending slab we can write the viscous resistance as the product of the pressure (τ) for the surface that is D so that

$$R = D\tau = \frac{2\mu v D}{D} = 2\mu v \tag{2.116}$$

The slab will move when he buoyancy overcome the resistance, that is

$$gDd\rho\alpha\Delta T \geq 2\mu v \rightarrow v = -\frac{gDd\rho\alpha T}{4\mu} \tag{2.117}$$

A dimensional analysis shows that d is the thickness of the so called boundary layer which is

$$d = \sqrt{\kappa t} = \sqrt{\kappa \frac{D}{v}} \tag{2.118}$$

So that finally the velocity

$$v = D\left(\frac{g\rho\alpha T\sqrt{\kappa}}{4\mu}\right)^{2/3} \tag{2.119}$$

We can now substitute (2.119) in (2.118) to eliminate v and obtain

$$\left(\frac{D}{d}\right)^3 = \frac{g\rho\alpha T D^3}{4\kappa\mu} \tag{2.120}$$

Because on the left of the (2.120) we have dimensionless quantity the same should be for the right hand side quantity which is the definition of the *Rayleigh Number*: for the convection to start in a fluid comprised between two surfaces at different temperature the Rayleigh number defined by

$$R_a = \frac{g\rho\alpha T D^3}{4\kappa\mu} \tag{2.121}$$

Must be larger that some critical value R_{ac}.

A more detailed theory fix this critical number around 1000 but in the mantle of the Earth could reach 3×19^6. The convective flux could be written as

$$q = \rho C_p v T = \rho C_p \kappa T \frac{D}{d^2} = \frac{KT}{D}\left(\frac{D}{d}\right)^2 = \frac{KT}{D} R_a^{2/3} \tag{2.122}$$

Now Davies uses another approach by defining the flux across the boundary layer as $q = KT/d$ and then defining the flux across the gap D as $q_k = KT/D$ then the Nusselt number becomes

$$N_u = \frac{q}{q_k} = \frac{qT}{KT} \approx R_a^{1/3} \tag{2.123}$$

References

Textbooks

Armitage, P.J.: Astrophysics of Planet Formation. Cambridge University Press (2010)
Armitage, P.J., Kley, W.: From Protoplanetary Disks to Planet Formation. Springer (2010)
Davies, G.F.: Dynamic Earth. Cambridge University Press (1999)
Kaula, W.M.: Introduction to Planetary Physics. Wiley (1968)
Safronov, V.S.: Evolution of the Protoplanetary Cloud and Formation of the Earth and the Planets. Nauka Press, Moscow (1969)

Articles

Abe, Y., Matsui, T.: The formation of an impact-generated H2O atmosphere and its implications for the early thermal history of the Earth. J. Geophys. Res. **90**, C535 (1995)
Armitage, P.J.: Lecture notes on the formation and early evolution of planetary systems, http://arxiv.org/abs/astro-ph/0701485v3 (2010)
Coradini, A., Federico, C., Lanciano, P.: Earth and Mars: early thermal profiles. Phys. Earth Planet. Inter. **31**, 145 (1983)
Davies, G.F.: Thermal histories of convective Earth models and constraints in radiogenic heat production in the Earth. J. Geophys. Res. **80**, 2517 (1980)
Elkins-Taton, L.T.: Magma oceans in the inner solar system. Annu. Rev. Earth Planet. Sci. **40**(1), 113 (2012)
Goldreich, P., Ward, W.R.: The formation of planetesimals. Ap. J. **183**, 1051 (1974)
Hayashi, C., Nakazawa, K., MIzuno, H.: Earth's melting due to the blanketing effect of the primordial dense atmosphere. Earth Planet. Sci. Lett. **43**, 22–28 (1979)
Ingersoll, A.P.: The runaway Greenhouse: a history of water on Venus. J. Atmos. Sci. **26**, 1191 (1969)
Kaula, W.M.: Thermal evolution of earth and moon growing by planetesimal impacts. J. Geophys. Res. **84**, 999 (1969)
Kaula, W.M.: The seven ages of a planet. Icarus **26**, 1–15 (1975)

Komabayashi, M.: Discrete equilibrium temperatures of a hypothetical planet with the atmosphere and the hydrosphere of one component-two phase system under constant solar radiation. J. Meteor. Soc. Japan **45**, 137 (1957)

Mac Donald, G.J.F.: Calculations on the thermal history of the earth. J. Geophys. Res. **64**(11), 1967 (1969)

Toomre, A.: On the gravitational stability of a disk of stars. Ap. J. **139**, 1217 (1964)

Zahnle, K.J., Kasting, J.F., Pollack, J.B.: Evolution of a steam atmosphere during Earth's accretion I. Icarus **74**, 62–97 (1988)

Chapter 3
The Chemistry of the Earth Origin

Now we understand that during its formation our planet got a steam atmosphere. Presumably during the cooling phase all the water vapor in the atmosphere would condense and form the ocean (we will talk about that later in the book). This steam atmosphere should have been substituted by a primordial atmosphere originated probably by the degassing of impact planetesinals. In this chapter we will review the evidence for the secondary origin of the atmosphere (it was not captured from the solar nebula) and some of the processes which determined the atmosphere composition.

3.1 The Noble Gas Inventory

At this point we can outline the possible history of our planet. As we have seen in the previous chapter the Earth was completed in about 50 million years and in the process got a core and a magma ocean that developed into a mantle. It is possible that the planetesimals also transported the components that were incorporated into the magma ocean and the mantle and among these components we would probably find also the noble gases. A further contribution to the atmospheric composition is the late veneer. A cartoon showing this possible history is shown in Fig. 3.1. In the planetologists mind, three elements formed the planets: rock, ice and gas. In this very simple scheme it is obvious to think that the composition of the terrestrial planets reflects in some way the environment in which they have formed. A proof of that could be found in the comparison of the abundance of the elements on the sun with the abundance found in a very special c1ass of meteorites called carbonaceous chondrites. Figure 3.2 shows the abundance of the heaviest noble gases relative to the Sun abundance. These data are the proofs that the atmosphere of the terrestrial planets is of a *secondary origin* that is it probably degassed from the solid Earth. As a matter of fact if it would have originated as a capture from the nebular atmosphere helium and hydrogen would have a much larger abundance. On the other hand as

G. Visconti, *Origin and Evolution of the Atmosphere and Ocean*,
https://doi.org/10.1007/978-3-032-02653-8_3

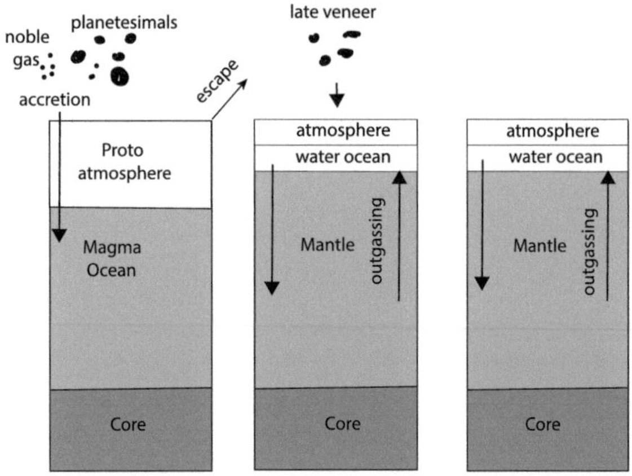

Fig. 3.1 The possible history of early Earth

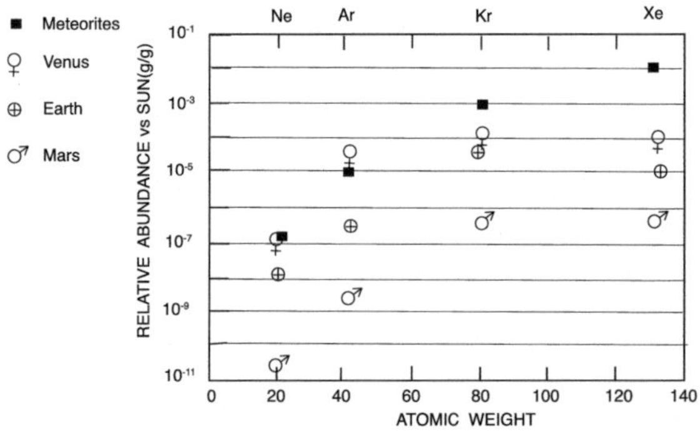

Fig. 3.2 The abundance of noble gases on the terrestrial planets

shown in Fig. 3.3 most of the elements are depleted with respect a solar atmosphere. We can define following Lodders and Fegley a depletion factor as

$$DF_{Ai} = \frac{(A_i/S_i)_{Earth}}{(A_i/S_i)_{Solar}} \tag{3.1}$$

From Fig. 3.3 we see that depletion factors are lower than 10^{-7} for noble gases Ne, Ar, Kr, Xe and the same is observed in Fig. 3.4. The fact that such elements show such large depletion is because they are inert while elements like carbon, etc. forms minerals and then can be preserved. The same behavior is detailed in Fig. 3.4 where a correlation is made between abundance in meteorites and the Earth. Except for some elements like beryllium, the figure shows a very good agreement between the two specimens. However the figure refers to elements with very low volatility, that is elements not found easily as a gas under normal conditions of temperature and pressure (on Earth). We need to think that, because the chondrites were processed through the same thermal cyc les as the material that makes up the planets, in the latter we could expect some depletion of volatiles. In the same Figure the same

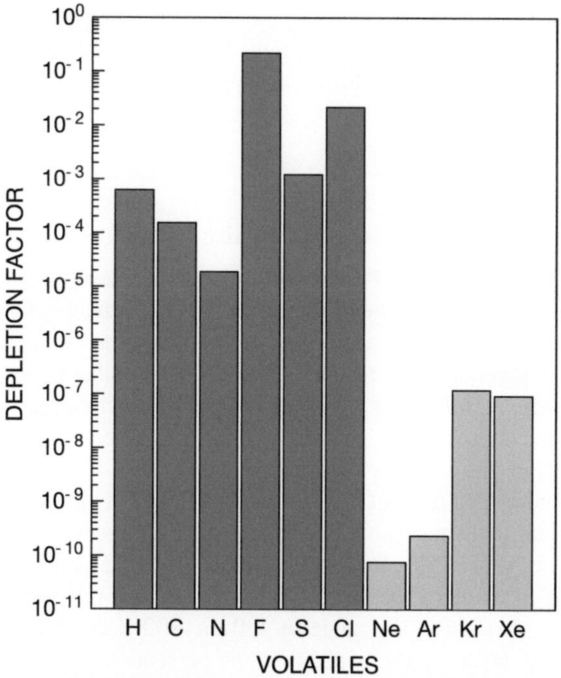

Fig. 3.3 The depletion factor for several volatiles

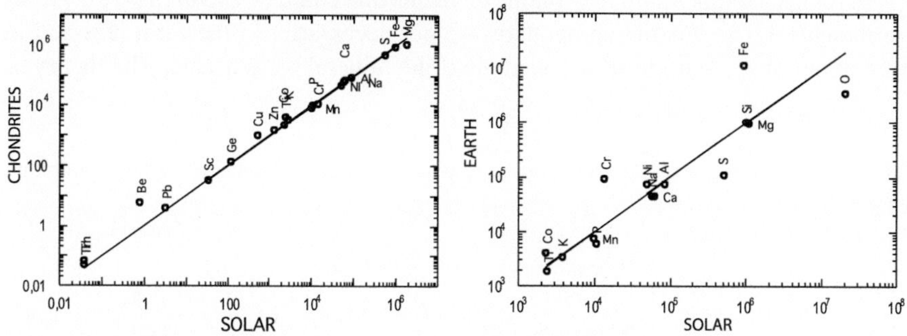

Fig. 3.4 The abundance of some elements in the solar atmosphere and meteorites (left) and solar and the Earth (right)

comparison for the abundance is made between the Sun and the Earth but this time we include some volatile element. In this case we note that elements like oxygen, sulfur, nitrogen, etc. deviate markedly from the strict proportions we have observed before. The straight line separates the depleted elements on the Earth (on the right) from the enriched ones (to the left). It is not completely correct to refer to elements that may have a very high chemical reactivity because their concentration may have also been determined by the chemical changes during the history of the Earth. A much safer attitude is to consider noble gases that, being chemically inert, may give a much more correct indication of volatiles retention by the planets. The volatile elements are in any case a very useful tracer in the history of the chemical evolution of a planet.

3.2 The Chemical Composition of the Planet

The easiest way to understand the differentiation in the nebula is to think about the formation of clouds. The water vapor, transported at some altitude, starts to condense when the temperature condition is such that its partial pressure is greater than saturation pressure. At this point the vapor is transformed into particles that can be either liquid or ice that show up as clouds or fog. In the nebula we have the same process although in this case the "vapors" are many, with different saturation pressures. We can understand how, in the central part of the nebula, those minerals with a high fusion point will condense while toward the edge we will have condensation of ices (not only of water). The atmosphere of the terrestrial planets originates, however, from gases like water vapor and carbon dioxide, and it is very hard to believe that materials condensing in the range of 400 to 1800 K could contain water. We could assume that condensation happens in the presence of chemical equilibrium. This implies that the process is very slow so that all the chemical reactions may go to completion. In this case it is possible to calculate the data reported in Table 3.1, where for each temperature the volatile elements that could be captured by the grains are indicated. The volatile compounds are not necessarily in the form reported in the column; they could form for example if the mineral were heated. The theory of

Table 3.1 The minerals in the nebula that may retain the indicated volatiles during equilibrium condensation

Equilibrium temperature (K)	Volatile bearing phase	Potential volatile(s)
1825	U,Th, in refractory Ca-bearing refractories	^{4} He
1520	C, N, Fe-Ni alloy	CO, CO_2, CH_4
1225	Scheibersite (Fe,Ni),P	Px PO_3, PH_3
950–1050	Feldspars, feldspathoids $(Na, K)AlSi_3O_8(Na, K)AlSi_3O_8$	^{40} Ar
895	Sodalite (3NaAlSiO_4NaCl)	Cl_2 HCl
766	Fluorapatite ($Ca_5(P0_4)$, F)	HF
714	Whitloekite(Ca, $(P0_4)$,)	P_x, PO_x,PH_3
687	Troilite (FeS)	S_x H_2S, COS, SO,
480	Tremolite(Ca, Mg, Si_8 O_{32}, $(OH)_2$	H, o, H" 0
460	Hydroxyapatite(Ca, $(P0_4)30H$)	H_2O, H_20_2
400	Talc((Mg, Fe), Si_3O_3, $(OH)_4$)	H_2O, H_20_2
400	Serpentine ((Mg, Fe), $Si_30_5(OH)_4$)	H_2O, H_20_2, O_3

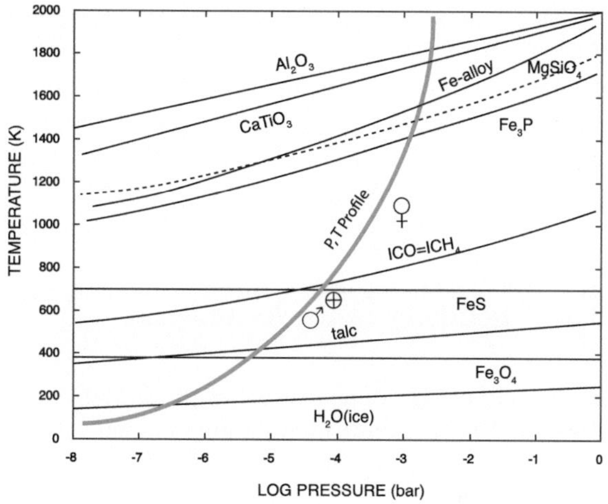

Fig. 3.5 Limit of stability of the minerals condensed in the primitive nebula. That is the curves indicate the saturation pressure for the vapors. The thick line is the politropic with the Nebula. Also shown are the approximate position of the terrestrial planets

equilibrium condensation is very much criticized but it is instructive in any case. Non equilibrium processes may happen when the gas cools in a time much shorter than the lifetime of reactions that happen in solid, gaseous or heterogeneous (solid and gas) phases. We will analyze these processes in some detail because they may imply important differences in the chemical composition of the terrestrial planets. Figure 3.5 shows the stability limit for different minerals (condensates) in the primitive solar nebula. These are essentially the saturation curves and no condensate is stable above the curve labeled W, which represents the stability curve of very high temperature refractories. Also shown is the adiabatic relation that we have mentioned before. From this figure it is quite evident how the terrestrial planets formed in a region where there was a large abundance of silicates and ferrous materials, while the Jovian planets formed where ices predominated. The condensation temperature is reported for a number of minerals together with the volatiles that the same minerals should contain. To the general rule high temperature—low volatile content, there are some exceptions, not worth discussing at the moment. From the same table it is possible to see (although complicated calculation is needed) that the Earth formed at lower temperature than Venus. For this reason it is enriched in hydrates (like tremolite, serpentine and talc). Mars formed in an even colder region containing a larger fraction of water. This table shows that according to the hypothesis of equilibrium condensation, planets behind Venus may contain very important volatile compounds like water vapor. Much more detailed calculation show, that it is possible to evaluate the abundance of the different volatiles on the planets. These calculations show that Venus formed with a water abundance much less than the Earth, and our planet has less water than Mars. These conclusions can be easily criticized by invoking the non-equilibrium condensation. Sometime these debates involve not only only science but also personal views that have nothing to do with scientific proofs. John Lewis has been a pioneer in these problems and a reader who wants to know more may read his nice book on the chemistry of the solar system. What matters to us is to have shown that in principle the terrestrial planets form with a certain content of volatiles. These are then degassed by a number of processes that may include volcanism after the planet formation or the impact of planetesimals during the planet formation. At this point we need to understand how the atmospheres differentiated and arrived at the composition that we know today. Before that we like to be more detailed on the chemistry inspired by a paper by Bruce Fegley.

3.3 Chemistry of Rocks and Gas

There is an old but quite instructive paper by Bruce Fegley and Ronald Prinn published in 1989 which fixes what kind of chemistry takes place in the nebula. This is done by comparing the energy flux Φ available for the reaction with the thermal energy flux $\Phi_T = \sigma T_e^4$ where σ is the Stefan Boltzmann constant and T_e the effective temperature of the nebula $\approx$300 K. Representative values for the relevant quantities are at 1 A.U., T, local temperature 600 K, E, activation energy for the reaction 5×10^4 J/mole, ϵ, the efficiency of conversion of convective energy in lightning strike $4 \times 10^{-6.}$ Then we have

$$\Phi(\text{thermochemistry})/\Phi_T \approx \exp(-E/kNT) = 4 \times 10^{-5} \tag{3.2}$$

With k Boltzmann constant and N Avogadro number

$$\Phi(\text{thunder-shock chemistry})/\Phi_T = \epsilon \approx 4 \times 10^{-6} \tag{3.3}$$

$$\Phi(\text{solar driven-photochemistry})/\Phi_T \approx 10^4 \times F \exp -\tau_r/\Phi_T \approx 0 \tag{3.4}$$

Where is the photon Flux $\approx 10^{-3}$ watt m^{-2}. Because τ_r is quite large. Also

$$\Phi(\text{solar driven-photochemistry}/\Phi_T \approx 10^{-9} \tag{3.5}$$

And finally

$$\Phi(\text{radiochemistry})/\Phi_T \approx 10^{-12} \tag{3.6}$$

From these simple considerations we deduce that the thermochemistry is the most important process in the nebula. It becomes even more important when we move toward the center where temperature approach 1500 K so that the ratio to Φ_T goes below 0.02. It is interesting to apply the above criteria to the present atmosphere with $T_e \approx 255$ and $T \approx 300$ K we see that the dominant process is the solar driven photochemistry (Notice that now T is no longer 600 K). We can refer to Fig. 3.5 where the condensation temperatures are shown as a function of pressure. Also shown is a politropic (p,T) curve that gives the approximate behavior of these quantities in the central plane of the nebula. For condensation temperature we intend when gas begins to change phase. Also shown are the approximate location of the terrestrial planets.

So the assumption in this case that the Earth at about 1 A.U. experience a temperature of roughly 600 K and a pressure of roughly 10^{-4} bar. Going further away at about 5 A.U. (roughly where now is the Jupiter orbit) we find the *snowline* where water vapor condenses to water ice at temperature of $\approx$150 K. This short paragraph will give a picture of the main chemical interactions of other key elements like Mg, Si, Fe and S that when combined in various way with oxygen constitute about 91% by mass of the material accreted by the terrestrial planets. The most abundant element which condense as shown in Fig. 3.5 as gaseous iron labeled as Fe. As the nebula cools the Fe alloy could react with phosphorous formind Fe_3P or the mineral schreibersite. At lower temperature ($\approx$700 K) the Fe ay recat with H_2S forming troilite FeS according

$$Fe \text{ (alloy)} + H_2S\text{(gas)} \rightarrow FeS\text{(gas)} + H_2\text{(gas)} \tag{3.7}$$

This consumes about 50% of iron. With further cooling the metallic iron combines with magnesium silicate to form forsterite and faialite

$$\begin{gathered} 2Fe\text{(alloy)} + 2MgSiO_3\text{(enstatite)} + 2H_2O\text{(gas)} \rightarrow \\ Mg_2SiO_4\text{(forsterite)} + Fe_2SiO_4 + 2FeSiO_4\text{(fayalite)} + 2H_2\text{(gas)} \end{gathered} \tag{3.8}$$

This reaction consumes all the remaining iron and the residual can reacts further with the water vapor in the nebula at about 370 K according to

$$3Fe \text{ (alloy)} + 4H_2O \text{ (gas)} \rightarrow Fe_3O_4\text{(magnetite)} + 4H_2\text{(gas)} \tag{3.9}$$

It is instructive also the study of the formation of other rocks material through reactions of magnesium with silicate

$$\begin{gathered} 2Mg \text{ (gas)} + SiO \text{ (gas)} + 3H_2O \text{ (gas)} \rightarrow \\ Mg_2SiO_4 \text{ (forsterite)} + 3H_2 \text{ (gas)} \end{gathered} \tag{3.10}$$

Enstatite forms at lower temperature

$$\begin{gathered} Mg_2SiO_4 \text{ (forsterite)} + SiO \text{ (gas)} + H_2O \text{ (gas)} \rightarrow \\ 2Mg_2SiO_3\text{(enstatite)} + H_2 \text{ (gas)} \end{gathered} \tag{3.11}$$

Some enstatite is consumed by reaction (3.8) bur remain stable at lower temperature where can forms serpentine

$$\begin{gathered} Mg_2SiO_4 \text{ (forsterite)} + Mg_2SiO_3\text{(enstatite)} + H_2O \text{ (gas)} \rightarrow \\ Mg_3Si_2O_5(OH)_4\text{(chrysotile serpentine)} \end{gathered} \tag{3.12}$$

This reaction is very slow while enstatite may be consumed with formation of talc

$$4Mg_2SiO_3(\text{enstatite}) + 2H_2O\ (\text{gas}) \rightarrow \\ Mg_3Si_4O_{10}(OH)_2(\text{talc}) + Mg(OH)_2(brucite) \tag{3.13}$$

All at temperature between 200 and 400 K. At this point we can make a short resume of the history of the solar system. After 1–2 million years the accumulation of the sun is completed as well as the formation of the terrestrial planets. it is clear that the material differentiated according to the sequence we have outlined which schematically illustrated Fig. 3.5, Up to 5–7 A.U. the building material for the planets are high condensation temperature minerals while beyond this distance are mostly ice materials a gases. However even within the terrestrial planets zone there is a different composition with water increasing from Venus to Mars. Starting from the Sun completion the star undergoes a T-Tauri star phase (see appendix) which generates strong winds that removes most of the remaining gases in the nebula. This phase last for about 3×10^6 years, We have to examine what happened to the water present in the atmosphere or degassed from the interior. A very important role in this respect was the escape of gases from the upper atmosphere.

3.4 Escape Mechanisms

The composition of formed atmosphere was subject to changes due to he interaction with the surface of the planet but also for some kind of evaporation when molecules in the tail of Maxwell distribution reach escape velocity. This being more likely for lighter gases like hydrogen or helium. For heavier gases (noble gases for example) they may be dragged by the mass movement of the atmosphere during hydrodynamic escape.

3.4.1 Jeans Escape

The molecular velocities in a gas are distributed according to the kinetic energy and according to the Boltzmann statistics. The distribution is known as Maxwellian and states that the number of molecules with velocities in the range v and v + dv is given by the function

$$f(v) = \frac{dn}{dv} = \frac{4n_0}{\sqrt{\pi}}\left(\frac{m}{2kT}\right)^{3/2} v^2 \exp\left(\frac{-mv^2}{2kT}\right) \tag{3.14}$$

In this equation n_0 is the number of molecules per unit volume, k the Boltzmann constant and m the mass of a molecule. This distribution resembles a bell curve, so that on the high velocity tail there is a number of molecules with velocities higher than the escape velocity $v_e = GM/r$ for a planet of mass M and at distance r

from the center. Not all these molecules will be able to escape because they may experience collisions or simply because their orientation is not right, that is they could be streaming downward and not in the upward hemisphere. We can take into account collisions by writing that the probability of collision for a molecule with geometric cross section σ is given by

$$\text{probability of collision} = \exp[\sigma N(r)] \tag{3.15}$$

where $N(r)$ is the molecular columnar density as a function of distance r. We may assume that the level at which the escape is possible (very low probability for collision) corresponds to the condition that the mean free path becomes equal to the scale height H

$$H = \frac{1}{n_e \sigma} \rightarrow n_e \sigma H = 1 \tag{3.16}$$

The altitude at which Eq. (3.16) is satisfied is called *exobase* and it is assumed that $N(r) = n_e N$. To understand the significance of H we notice that number density n changes according to the Boltzmann law

$$n = n_0 \exp -\frac{\Delta\phi}{kT} \tag{3.17}$$

where $\Delta\phi$ is the difference in potential energy

$$\Delta\phi = GMm\left(\frac{1}{r_0} - \frac{1}{r}\right) \tag{3.18}$$

We notice that the quantity

$$\lambda = \frac{GMm}{kTr} = \left(\frac{v_{esc}}{U}\right)^2 = \frac{r}{H} \tag{3.19}$$

Is the ratio of the escape velocity $v_{esc}^2 = 2\text{GMm}/\text{r}$ to the most probable thermal velocity $U = (2kT/m)^{1/2}$ while $H = kT/(mg)$ is the local scale height. We have for a temperature of 500 K a scale height of about 425 km for hydrogen. A typical value of $\sigma \approx 10^{-15}\text{cm}^{-2}$ so that $n_e \approx 2.4 \times 10^7\ \text{cm}^{-3}$. Then (3.17) becomes

$$n = n_0 e^{\lambda - \lambda_0} \tag{3.20}$$

To evaluate the escape flux we consider situation like Fig. 3.6 where the vertical component of the velocity is $vcos\theta$ that must be integrated alk over the solid angle

$$\Phi_J = \int_0^{2\pi} d\phi \int_0^{\pi/2} d\theta \int_0^{\infty} f(v) \cos\theta \sin\theta dv \tag{3.21}$$

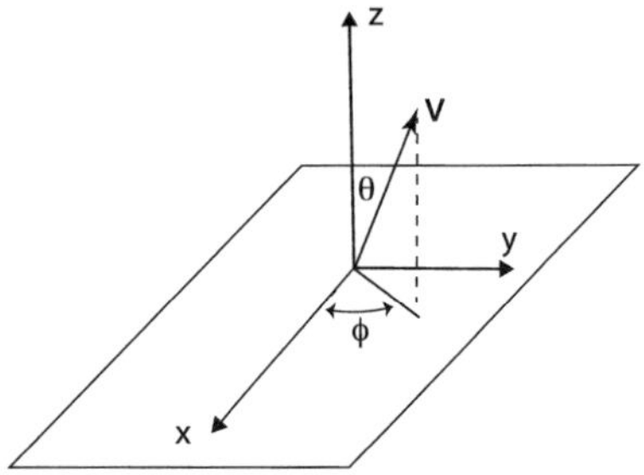

Fig. 3.6 The geometry of molecular escape

Solving the integral we have

$$\Phi_J = \frac{U}{2\sqrt{\pi}}(1 + \lambda_c)e^{\lambda_e} = n_c w_e \tag{3.22}$$

Where w_e is the effusion velocity. Once we have determined the escape flux it is very instructive to find how long it would takes to lose the molecules contained in an atmospheric column. This time τ_e can be obtained by dividing the density above the exobase ($n_e H_e$) by the escape flux so that (keep in mind the definition for λ).We have

$$\tau_e = \sqrt{\pi}\frac{U}{g}\frac{e^{\lambda_0}}{(1 + \lambda_0)} \tag{3.23}$$

where the flux is evaluated at $r = r_0$ the radius of the planet. There are few objections in using this equation because does not take into account that temperature change with height, A better definition of escape time could be based on the loss of planetary mass M

$$\tau_e = \left(-\frac{1}{M}\frac{dM}{dt}\right)^{-1} = \frac{Q}{4\pi G}\sqrt{2\pi kTm}^{-3/2}\frac{e^{\lambda_c}}{1 + \lambda_c} \tag{3.24}$$

Order of magnitude arguments show that evaporative escape happens when the thermal energy of a single molecule $3kT/2$ is comparable to the potential energy of the molecule so that according (3.19) a reasonable value for $\lambda \approx 1.5$. Actually we can assume $\lambda \approx 2$ which corresponds to the so called *blowoff* or hydrodynamic escape when the atmosphere as a whole expands roughly at the speed of sound c_s. The solutions (3.23)–(3.24) give quite different values because for T = 1000 K the first gives roughly 1 minute while the second gives 10^{14} years so that the entire life of the universe would not be enough to dissipate the atmosphere. As a preliminary guess we may assume that the escape flux it is just the product of the density n_s at the sonic level that is when $U =\approx c_s$, $\phi_s = n_s c_s \approx n_s U$. Considering the spherical geometry the flux can also be written

$$\phi_s(r_0) \approx \left(\frac{r_s}{r_0}\right)^2 n_s U = \left(\frac{r_s}{r_0}\right)^2 n_0 e^{\lambda_s - \lambda_0} \tag{3.25}$$

3.4.2 The Limiting Flow

We have found expression for the evaporation flux but the problem is that there are the conditions to feed such flow. We show in the appendix that the total flow of a i component is given by

$$\phi_i = -D_i n_i\left(\frac{1}{n_i}\frac{dn_i}{dz}+\frac{m_i g}{kT}+\frac{1+\alpha}{T}\frac{dT}{dz}\right)-Kn_i\left(\frac{1}{n_i}\frac{dn_i}{dz}+\frac{m_a g}{kT}+\frac{1}{T}\frac{dT}{dz}\right) \tag{3.26}$$

In this case D_i is the molecular diffusion coefficient and K the eddy diffusion coefficient and α. the thermal expansion. m_i is the molecular mass of the component and m_a that of the atmosphere. D_i can be written as $D_i = b_i/na$ where b_i is the binary collision parameter. At this point we consider a component n_1 immersed in a component with density n_2 and define the mixing ratio f_1

$$f_1 = \frac{n_1}{n_2} \tag{3.27}$$

Then it is found

$$\frac{1}{H_1} = -\frac{1}{n_1}\frac{dn_1}{dz} = \frac{1}{H_f}+\frac{m_2 g}{kT}+\frac{1}{T}\frac{dT}{dz} \tag{3.28}$$

Equation (3.26) can be rewritten with b_1 substituting $D(n_1+n_2)$

$$\phi_1 = \phi_l - (Kn_2+b_1)\frac{df_1}{dz} = \phi_l - (K+D)n_2\frac{df_1}{dz} \tag{3.29}$$

Where

$$\phi_l = b_1 f_1\left(\frac{1}{H_2}-\frac{m_1 g}{kT}-\frac{1+\alpha}{T}\frac{dT}{dz}\right) = \frac{b_1 f_1}{H'} \approx \frac{f_1}{H_2}\left(1-\frac{m_1}{m_2}\right) \tag{3.30}$$

According to (3.29) there must be a gradient in the mixing ratio unless $\phi_i = \phi_l$. If $\phi_i > \phi_l$ the the gradient of the mixing ratio must be negative and there will be level where the flux will be chocked off. There is a further simplification for low mixing ratio (like is the case for H and H_2 and that is

$$\phi_l \approx \frac{b_1 f_1}{H_2} \tag{3.31}$$

We will give some number in a while after we have illustrated the blow off mechanism

3.4.3 The Hydrodynamic Escape

We have already mentioned the existence of an escape mechanism that involves heavier molecules than hydrogen and helium. In this case there is some kind of expansion of the whole atmosphere of the planet and the lighter gases drag the heavier ones. We have also seen that this situation arises when the thermal velocity U is near to the speed of sound and $\lambda = 2$. There is nothing dramatic about that because with the increase of the mean free path with altitude the velocity of molecule also increases reaching the speed of sound. Also the physics does not change and a pioneer as James Walker suggested many years ago that if the sonic level is higher than the exobase the description is that of Jeans otherwise we must use the blow off. A more clear statement about the two kind of escape is that in the case of Jeans the descrip tion is based on the kinetic theory of gas (i.e. the maxwellian distribution) and the fluxes are low while in the hydrodynamic case we need to use the continuum model (i.e. the fluid dynamic laws), In this case the mean free path is small with respect to the scale height, We then assume to have a gas 1 that diffuses into a gas 2 and will denote with F_I and F_2 the corresponding fluxes. The relative velocities will be given by

$$\begin{aligned} w_1 - w_2 &= \frac{F_1}{n_1} - \frac{F_2}{n_2} = \frac{b}{n_2}\left(\frac{1}{n_1}\frac{dn_1}{dr} + \frac{m_1 g}{kT}\right) \\ w_2 - w_1 &= \frac{F_2}{n_2} - \frac{F_1}{n_1} = \frac{b}{n_1}\left(\frac{1}{n_2}\frac{dn_2}{dr} + \frac{m_2 g}{kT}\right) \end{aligned} \tag{3.32}$$

In these equations r indicates the distance from the center of the Earth and replaces the altitude. The second equation gives the velocity of gas 2 that diffuses in gas 1. From these equations it is possible to obtain the gradients

$$\frac{dn_1}{dr} = -\frac{m_1 g}{KT}n_1 + \frac{1}{b}(n_1 F_2 - n_2 F_1), \qquad \frac{dn_2}{dr} = -\frac{m_2 g}{KT}n_2 + \frac{1}{b}(n_2 F_1 - n_1 F_2) \tag{3.33}$$

Adding we get

$$\frac{dn_1}{dr} + \frac{dn_2}{dr} = \frac{dn_a}{dr} = (n_1 m_1 + n_2 m_2)\frac{g}{kT} \tag{3.34}$$

This relation gives the variation of the total density $n_a = (n_1 + n_2)$ with radius. We define the mole fraction

$$X_2 = \frac{n_2}{n_1 + n_2} = \frac{n_2}{n_a} \tag{3.35}$$

So we have

$$\frac{1}{X_2}\frac{dX_2}{dr} = \frac{1}{n_2}\frac{dn_2}{dr} - \frac{1}{n_a}\frac{dn_a}{dr} \tag{3.36}$$

Substituting (3.33) for the first r.h.s. term and (3.34) for the second we have

$$\frac{1}{X_2}\frac{dX_2}{dr} = -\frac{m_2 g}{kT} + \frac{1}{b}\left(F_1 - \frac{X_1}{X_2}F_2\right)$$
$$+(m_1 X_1 + m_2 X_2)\frac{g}{kT} = \frac{1}{b}\left(F_1 - \frac{X_1}{X_2}F_2\right) - (m_2 - m_1)\frac{X_1 g}{kT} \tag{3.37}$$

Where we have used the fact that $X_2 = 1 - X_1$. Conservation of matter requires that the fluxes, F_1 and F_2, vary inversely as the square of the distance from the center of the planet. Gravitational acceleration also varies inversely as the square of r. The equation for the height variation of mole fraction therefore becomes

$$\frac{1}{X_2}\frac{dX_2}{dr} = \left[\frac{1}{b}\left(F_1 - \frac{X_1}{X_2}F_2\right) - (m_2 - m_1)\frac{X_1 g}{kT}\right]\left(\frac{r_0}{r}\right) \tag{3.38}$$

Where the values of the fluxes and gravity refer to some level r_0. This relation could be integrated but a solution could also be obtained by setting the right hand side to zero. In this case both X_1 and X_2 are constant and we get for the flux

$$F_2 = \frac{X_2}{X_1}F_1\left[1 - \frac{bgX_1}{kTF_1}(m_2 - m_1)\right] \tag{3.39}$$

This relation has a very simple physical interpretation. We may assume to have a gas with molecular mass m_2 mixed with a gas with molecular mass m_1. A molecule of the first gas will be subjected to the gravity $m_2 g$, to the buoyancy force in the opposite direction $m_1 g$ and to a viscous force proportional to the difference between the velocities of the molecules $(w_1 - w_2)$. The proportionality coefficient can be found to be kT/b, and because of the relation between flux and velocity the resistant force will be

$$\frac{kT}{b}\left(\frac{F_1}{X_1} - \frac{F_2}{X_2}\right)$$

At the equilibriun we have

$$(m_2 - m_1)g = \frac{kT}{b}\left(\frac{F_1}{X_1} - \frac{F_2}{X_2}\right) \tag{3.40}$$

From which we can obtain (3.39). It is useful to introduce a cross over mass m_c such that

$$m_c = m_1 + \frac{kTF_1}{bgX_1} \tag{3.41}$$

As a mass for the heavier component. For such a mass $F_2 = 0$ and (3.39) becomes

$$F_2 = \frac{X_2}{X_1}F_1\left(\frac{m_c - m_2}{m_c - m_1}\right) \tag{3.42}$$

For masses smaller than the separation mass, the lighter gas drags the heavier gas, while for masses larger than m_c the flux is zero because the buoyancy force is not enough to compensate for the gravity and viscous force. Using Eq. (3.41) we can ca1culate the hydrogen fluxes required to drag heavier molecules. The energy source for the flux of the lighter gas is the absorption of solar energy. Then, if S is the absorbed solar flux and F_1 is the molecular flux and each molecule has the gravitational energy GMm/r , we have

$$F_1 = \frac{Sr}{GMm} \tag{3.43}$$

That substituted in (3.41)

$$m_c - m_1 = \frac{kT}{bX_1} \frac{(Sr^3}{(Gm)^2} m \tag{3.44}$$

Using this relation we can build Table 3.1 with the limit flux and the critical or separation mass for typical bodies of the solar system. We notice that the present EUV (extreme ultraviolet) flux on the Earth is about half that given in the table. Even in this case atomic hydrogen could only drag molecules a little bit heavier than molecular hydrogen. In the past, however, the EUV flux could have been much higher than the present one, and thus it could be reasonable to expect the escape of much heavier gas. To understand how the atmosphere could be depleted of a certain gas as a function of a lighter gas, we can use a very simple theory. We assume that the fractional changes in the concentrations of the two gases is given by the relation

$$\frac{dN_1}{N_1} = (1+y)\frac{dN_2}{N_2} \tag{3.45}$$

where $(1 + y)$ is the fractionating factor and in general is slightly larger than 1. The depletion factor may be found by substituting Eqs. (3.41) in (3.39), where X_1/X_2, is substituted with N_1/N_2. We can find a relation between flux and mass of separation

$$F_2 = \frac{N_2}{N_1} F_1 \left(\frac{m_c - m_2}{m_c - m_1)} \right) \tag{3.46}$$

Based on this we can easily find the change with time of the gas being dragged

$$\frac{1}{N_2}\frac{dN_2}{dt} = -\frac{N_2}{F_2} = -\frac{F_1}{N_1}\left(1 - \frac{m_2 - m_1}{m_c - m_1}\right) \tag{3.47}$$

Comparing (3.46) to (3.45) m we have

$$1 + y = \frac{m_c - m_1}{m_c - m_2} \tag{3.48}$$

So that the solution to (3.47) becomes (Table 3.2)

Table 3.2 Escape fluxes in atomic mass units (a.m.u.) $m^{-2}s^{-1}$ for an absorbed solar flux of $10^3\,jcm^{-2}s^{-1}$. Separation mass m_c are also indicated. For the planetesimal the same density of the moon has been assumed and a radius of 1000 km

	Earth	Mars	Moon	Planetesimal
Flux	10^{16}	5×10^{16}	2×10^{17}	5×10^{17}
Mass (m_c)	2.5	20	197	1030

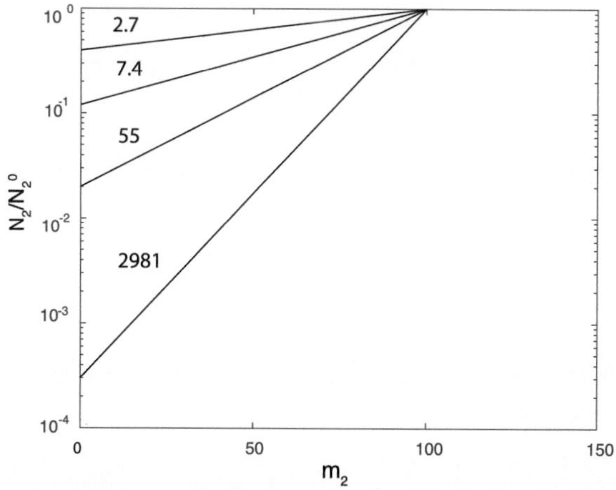

Fig. 3.7 Evolution of the component N_2 as a function of mass m_2. The number above the lines is the initial ratio N_1/N_1^0

$$N_2 = N_2^0 \left(\frac{N_1}{N_1^0}\right)^{1/(1+y)} \tag{3.49}$$

with (0) indicating the initial value. With the evolving composition it is expected that also the separation mass may change, so that y may be a function of time, and Eq. (3.49) may be actually an oversimplification. In order to understand how a solution of this kind works, we will take the most simple case, when the separation mass is constant. In this case the solutions, shown in Fig. 3.7, are determined by the ratio between the initial and final abundance of the lightest gas. If is that ratio that defines the slopes of the lines.

Keeping the separation mass constant and for an assigned heavier gas, the solutions show that the depletion of the latter increases with the depletion of the lighter gas. The solutions on a logarithm scale are straight lines originating at m_c (in this case the flux is zero).

A real curve for fractionation probably is not a straight line in this coordinate system. We could expect that as the solar flux decreases the separation mass also decreases and then the heaviest gases remains constant with time, so that the slope of the line will also decrease with time. The result for a real episode is that the fractionation curve is concave toward the negative of the abscissa. Based on what we have said, it looks like hydrodynamic escape is a mechanism capable of depleting the atmosphere of the heavy elements. The problems remaining are a UV flux two orders of magnitude larger than the present one, and a quantity of hydrogen large enough to be the carrier gas during the blow-off episodes. The astrophysicists that have invented the so—called T-Tauri phase for the stars may solve the first difficulty. Just after their birth, stars might have a period when their luminosity in the ultraviolet below 100 nm seems to be adequate. For hydrogen, the only way to get it is the photodissociation of water vapor. This as we have seen probably was present in large quantities on the Earth and the terrestrial planets at the beginning of the evolution of their atmospheres.

3.4.4 Other Escape Mechanisms

Donald Hunten has published a paper in 1982 in which he discusses in detail the various escape mechanisms which are known as non thermal. Since then the situation has remained the same and can be summarized as follows

Charge Exchange

In this process an ion with excess energy converts into an atom with the same energy: for example

$$H + H^{+*} \rightarrow H^{+} + H^{*}$$
$$O + H^{+*} \rightarrow O^{+} + H^{*}$$

The protons in the present Earth are taken from the *plasmasphere* wíth temperature in the range 2000–5000 K. This process accounts for 60–90% of the loss of hydrogen from the Earth.

Dissociative Recombination
In this case a fast ion is neutralized and transmits energy to one or both components

$$O_2^{*} + e \rightarrow O^{*} + O^{*}$$
$$OH^{*} + e \rightarrow O^{*} + H^{*}$$

Impact Dissociation and Photodissociation
Here excited atoms may be produced either with impacting electrons or the interaction with ultraviolet (UV) radiation

$$N_2 + e \rightarrow N^{*} + N^{*}$$
$$O_2 + h\nu \rightarrow O^{*} + O^{*}$$

Such reactions may produce energy in the range of a few eV not enough to energize hydrogen atoms that remain confined in a "corona" around the exobase over the pole. There are other processes like *sputtering, ion escape* which have however a limited influence.

3.5 Some Conclusion

In this chapter we learned that the composition of terrestrial planet reflected the environmental condition of the nebula that is minerals in the inner planets accumulated following their point of fusion. Accordingly for example Venus contains much less water than the Earth which has less water than Mars. The primordial atmosphere however did not reflected the nebula composition but had a secondary origin. This conclusion may reached comparing the rare gas content in meteorite and the Earth and Sun atmospheres. The latter being essentially that of the nebular gas. The atmosphere of the Earth probably originated from degassing of the accumulated materials and much of the primarily atmosphere (mostly hydrogen and helium) was lost to Jeans and hydrodynamic escape. At the present time the hydrogen flux is 2.5×10^8 cm^{-2} s^{-1}. This flux could not be very much different also in ancient times because of the low density in the originating regions and the lower luminosity of the Sun. A much more efficient way to dissipate the primary atmosphere is the hydrodynamic escape. In this case the mass motion of the atmosphere could have dragged H, H_2 and even methane CH_4 but also heavier atoms like those of rare gases. The hydrogen gas could results from the photodissociation of the water. A possible proof of the existence of this mechanism can be found in the abundance of the rare gases that shows lighter gas much more depleted as it is shown in Fig. 3.2 and discussed several times.

The Earth history however present violent episodes which included bombardment during the accretion phase but also large single impact like the formation of the moon. Some of these episodes could have provided gases to the atmosphere but also wiped out large portion of the existing atmosphere as we shown in the appendix. When the radius of the impactor is much larger than an atmospheric scale height and its velocity much larger than the planetary escape velocity, some of the planetary atmosphere may be driven off into space.

We are now on the verge of discuss the origin and the chemistry of the primitive atmosphere.

Appendix

Appendix Dissipation by the T-Tauri Wind

The T-Tauri winds is manifested in very young stellar object and consists in a bipolar outflow of gases of the order of 10^{-8} solar masses/year. This wind could have contributed to "clean" the space after the planet formation. One of the first treatment of this phenomena was accomplished by Minuro Sekiya in a paper published almost forty five years ago. Consider that based on solar abundance, the initial mass of rare gas in the Earth's atmosphere is roughly 2×10^{20} kg and now is reduced to 3×10^{14} kg. The work of Sekiya conside that the nebular gas consisted in H_2 and

H_e and that the flow was essentially radial and only the hydrogen flow is considered. We evaluate now the difference in velocity of the H_2 molecules u_0 and of the rare gas u_i

$$\Delta u_i = u_0 - u_i << u_0 \tag{3.50}$$

If this condition is satisfied the chemical homogeneity s retained and the rare gas are almost completely dragged. We now consider a dominant gas with number density n_0 and molecular mass m_0 while the rare gas has mass m_i and number density n_i. It can be shown that the difference (1) is a function of the radial pressure gradient

$$\Delta u_i \approx -\frac{m_i}{m_0} D_i \frac{1}{p} \frac{\partial P}{\partial r} \tag{3.51}$$

where P is the pressure and D_i the diffusion coefficient. The pressure gradient can be expressed by the hydrostatic equilibrium

$$\frac{\partial P}{\partial r} = -\rho g = -nm_0 \frac{GM}{r^2}$$

And because $P = nkT$ we have

$$\frac{\partial}{\partial r} \ln P = -\frac{GM}{r^2} \frac{m_0}{kT} \tag{3.52}$$

Consider now that $D_i = f_i(T)/P$ and introduce the mass flux $\dot{M}_g = 4\pi r^2 \rho u_0 = 4\pi r^2 n m_0 u_0$ therefore (3.52) becomes

$$bALT \frac{\Delta u_i}{u_0} = \frac{4\pi GM m_0 m_i f_i(T)}{\dot{M}_g (kT)^2} \tag{3.53}$$

Now consider that the quantity $r^2 kT/GMm_i$ it is just the scale height

$$u_0 = \frac{\dot{M}_g}{4\pi r^2 \rho} \tag{3.54}$$

Then we have

$$\Delta u_i = \frac{GMm_i D_i}{kTr^2} \approx \frac{D_i}{H_{ei}} \tag{3.55}$$

We will show that this result is obtained from the diffusion equation. The condition for the escape of a rare gas is obtained by imposing that $\Delta u_i/u_0 \approx 1$. The values of $\dot{M}_g$ obtained are function of temperature and for example for Argon at 1000 K we get a value around 10^{13} kg/year. The case we have examined the hydrogen flux is the thermal escape flux, which apparently will be able to dissipate the rare gas in a reasonable time that is $2 \times 10^{20}/10^{13} \approx 10^7 \, years$. To get rid of the nebular gas it is necessary to recur to the T-Tauri wind. We will follow the simple treatment of an old paper by Horedt. We will show only that a T-Tauri wind is able to transport at great distances several time its initial mass. He considered a spherical stationary flow at distance r, with the gas moving with velocity v and density ρ. Conservation of momentum between r and $r + \Delta r$ requires

$$(m + \Delta m)(v + \Delta v) - mv = F\Delta t \tag{3.56}$$

with $\Delta r = v\Delta t$ and

$$F = \frac{GMm}{r^2} - \frac{m}{\rho}\frac{\Delta p}{\Delta r} \tag{3.57}$$

Substituting in (3.56) we have

$$v\Delta v + v^2\frac{\Delta m}{m} + GM\frac{\Delta m}{r^2} + \frac{\Delta p}{\rho} = 0 \tag{3.58}$$

The pressure can be expressed as a function of density ρ following the politropic $p = const\ \rho^\gamma$ and (3.58) easily integrated

$$\frac{v^2}{2} + \bar{v^2}\ln m - \frac{GM}{r} - \frac{\gamma}{\gamma - 1}\frac{RT}{\mu} = cost \tag{3.59}$$

Where μ is the molecular weight. The constant is determined by the boundary conditions at $r = R$ and $r \to \infty$

$$v = 0, \quad T = T_R, \quad m = \dot{M} = \frac{\Delta M}{\Delta t} \qquad \text{when r = R}$$

$$v = v_\infty, \quad T = 0 \quad m = m_\infty \qquad \text{when } r = \infty$$

The we get

$$\frac{v_\infty^2}{2} + \frac{GM}{R} - \frac{\gamma}{\gamma - 1}\frac{RT_R}{\mu} + \bar{v^2}\ln\frac{m_\infty}{\dot{M}} = 0 \tag{3.60}$$

If the mass transported for unit time is the same at all the distances then $m_\infty = \dot{M}$ and (3.60) becomes

$$v_{c\infty}^2 = 2\frac{GM}{R} - 2\frac{\gamma}{\gamma - 1}\frac{RT_R}{\mu} = 0 \tag{3.61}$$

Which is the velocity at infinity. Now the average can be calculated as

$$\bar{v^2} = \int_0^{v_{c\infty}} \frac{v^2}{v_{c\infty}} dv = \frac{{v_{c\infty}}^2}{3}$$

So that (3.60) becomes

$$v_{c\infty}^2 = 2\left(\frac{\gamma}{\gamma.1}\frac{RT_r}{\mu} - \frac{GM}{R}\right)\left(1 - \frac{2}{3}\ln\frac{m_{c\infty}^2}{\bar{M}}\right) \tag{3.62}$$

If $v_{c\infty}$ must be positive then $\ln(m_{c\infty}^2/\dot{M}) < 3/2$ and $m_{c\infty} < 4.5\dot{M}$ that is the wind is able to transport to great distance several times its initial mass.

More on Escape Flux

To set things straight on the escape flux we will follow an old paper by Donald Hunten that somehow simplifies the problems but is fundamentally correct. We start by defining the hydrostatic equilibrium in the presence of a non isothermal atmosphere.

$$dp = kTdn_i + kn_idT = -\rho g = -n_i m_i g \rightarrow \frac{dn_i}{dz} = -\frac{n_i}{H_i} - \frac{n_i}{T}\frac{dT}{dz} - \frac{\phi_i}{D_i} \tag{3.63}$$

Where $H_i = kT/M_i g$ and the last term is the contribution of the diffusive flux. The sum of molecular diffusion and eddy diffusion refers to a flux ϕ_1 of the component with number density n_1 which diffuses into a component n_2

$$\phi_1 = -D_1 n_1\left(\frac{1}{n_1}\frac{dn_1}{dz} + \frac{1}{H_{e1}}\right) - Kn_1\left(\frac{1}{n_1}\frac{dn_1}{dz} + \frac{1}{H_{e2}}\right) \tag{3.64}$$

Where n_1 is the number density of the minor component, D_1 it molecular diffusion coefficient, K the eddy diffusion coefficient while H_{e1} and H_{e2} are the scale heights

$$\frac{1}{H_{e1}} = \left(\frac{m_1 g}{kT} + \frac{1+\alpha_1}{T}\frac{dT}{dz}\right), \quad \frac{1}{H_{e2}} = \left(\frac{m_2 g}{kT} + \frac{1}{T}\frac{dT}{dz}\right) \tag{3.65}$$

The Eq. (3.63) has been derived by Chapman and Cowling in a classic book and can also be justified with a simple argument by introducing the volume mixing ratio $f_1 = n_1/n_2$ where now n_2 is the number density of the background gas. We approximate the flux to be proportional to the gradient of the mixing ratio df_1/dz

$$\phi_1 = -Kn_2\frac{df_1}{dz} = -Kn_2\frac{d}{dz}\left(\frac{n_1}{n_2}\right) = -Kn_1\left(\frac{1}{n_1}\frac{dn_1}{dz} - \frac{1}{n_2}\frac{dn_2}{dz}\right) \tag{3.66}$$

For an isothermal atmosphere the term $1/n_2(dn_2/dz)$ it is just the scale height $m_2 g/kT$. α_1 is the thermal diffusion coefficient. Based on (3.65) we may define a mixing ratio scale height as

$$\frac{1}{f_1}\frac{df_1}{dz} = -\frac{1}{n_1}\frac{dn_1}{dz} - \frac{1}{H_{e2}}, \quad \frac{1}{H_f} = \frac{1}{H_1} - \frac{1}{H_{e2}} \tag{3.67}$$

Equation (3.63) can now be written using $b_1 = D_1(n_1 + n_2)$

$$\phi_1 = \phi_l - (Kn_2 + b_1)\frac{df_1}{dz} \tag{3.68}$$

where

$$\phi_l = b_1 f_1 \left(\frac{1}{H_2} - \frac{1}{H_{e1}}\right) = \frac{b_1 f_1}{H'} \approx \frac{b_1 f_1}{H_2}\left(1 - \frac{m_1}{m_2}\right) \tag{3.69}$$

The *limiting flux* does not depend, on the distribution of component 1, but only on it value, the diffusion coefficient and its mass. If the mixing ratio stay constant with altitudei the flux ϕ_1 coincides with the limiting flux. In general we can say that below the homopause the dominant transport mechanism the eddy diffusion (Eq. 3.66) and it worth to work out a simple example where the component has a loss (with lifetime $1/J$. The continuity equation in the vertical direction gives

$$\frac{d\phi_i}{dz} = Jn_i \rightarrow -K\frac{d}{dz}\left(\frac{1}{n_i}\frac{dn_i}{dz} - \frac{1}{n_a}\frac{dn_a}{dz}\right) = Jn_i \rightarrow -K\frac{d}{dz}\left(-\frac{1}{H_i} + \frac{1}{H_a}\right) = Jn_i \tag{3.70}$$

where if we assume J to be constant and $n_i = n_{i0}e^{-z/H_i}$: Then (3.70) can be integrated to give

$$-K\left(-\frac{1}{H_i} + -\frac{1}{H_a}\right) = JH_i \rightarrow \frac{1}{H_i^2} - \frac{1}{H_i H_a} - \frac{J}{K} = 0 \tag{3.71}$$

And the solution for the scale height

$$\frac{2}{H_i} = \frac{1}{H_a} * \left(\frac{1}{H_a^2} + \frac{4J}{K}\right)^{1/2} \tag{3.72}$$

So that for large K the mixing is complete $1/H_i = 1/Ha$.

Impact Erosion of the Atmosphere

A catastrophic way to get rid of a good portion the atmosphere is through an impact with a large object. This could have happened about 4.5 billion years for the Earth when the moon formed or may have happened during the same epoch for the atmosphere of Mars. An the present time there exist a number of numerical simulation of this event that we cannot reproduce here so we will resume the approach that was popular at the end of the eighties. In particular we will refer to a paper published more than 30 years ago by Jay Meloshand Ann Vickery on the impact erosion of martian atmosphere. The process is schematically shown in Fig. 3.8, In the phase A a projectile enter the atmosphere compressing, accelerating and heating the gas ahead

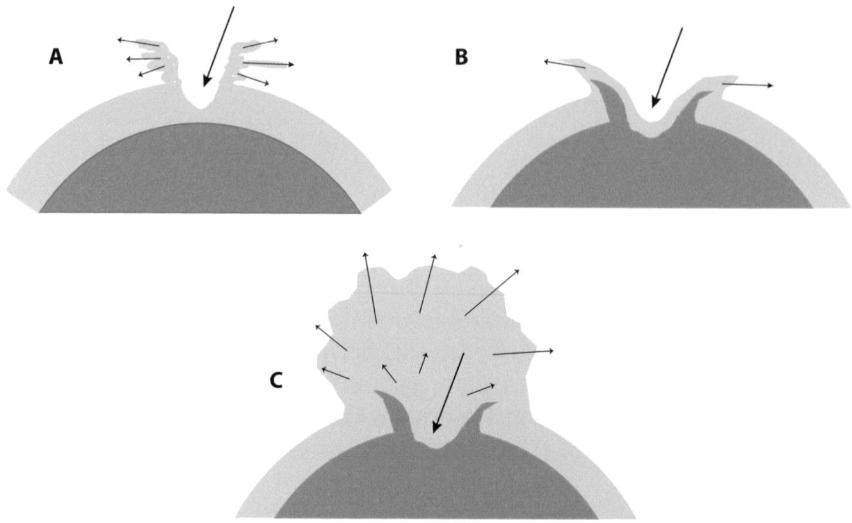

Fig. 3.8 Different phases of the collision of a large object with the Earth. In **A** the atmosphere is compressed and heated ahead of the projectile. In **B** the growing crater expels fragments in the atmosphere, while in **C** the expanding cloud reaches the upper atmosphere and its pressure equals that of the surroundings

of it. In B solid ejecta from the forming crater transfer part of their momentum to the atmosphere and finally in C the generated hot plume move outward and upward. For small projectiles the expansion stops when the pressure of the plume equals the surrounding atmospheric pressure although some buoyancy is generated because the plume is hotter than the surrounding. For more massive projectile the equilibrium is not reached equilibrium at radii less than the scale height. If we assume E to be the specific energy (J kg^{-1}) and H_{vap} the vaporization energy conservation gives the mean expansion velocity of the plume as

$$\tfrac{1}{2}v^2 = E - H_{vap} \tag{3.73}$$

H_{vap} is about 13×10^6J kg^{-1} *to establish* for silicates and about 3MJ/kg^{-1} for ice. The specific energy of the incoming projectile is $u^2/2$ if u is the gas velocity of the shock wave the maximum velocity $u = v/2$ where v is the impact velocity. So we arrive at $E = v^2/8$ and based on (3.73) the minimum threshold for the impact velocity for the plume to reach escape velocity is

$$v_{min} = \left[8\left(\frac{v_{esc}^2}{2} + H_{vap}\right)\right]^{1/2} \tag{3.74}$$

In the case of Mars it results that $v_{min} = 14.3$km s^{-1} for silicate and 11.3km s^{-1} for ice and these velocity are less than most Earth crossing asteroids or comets.

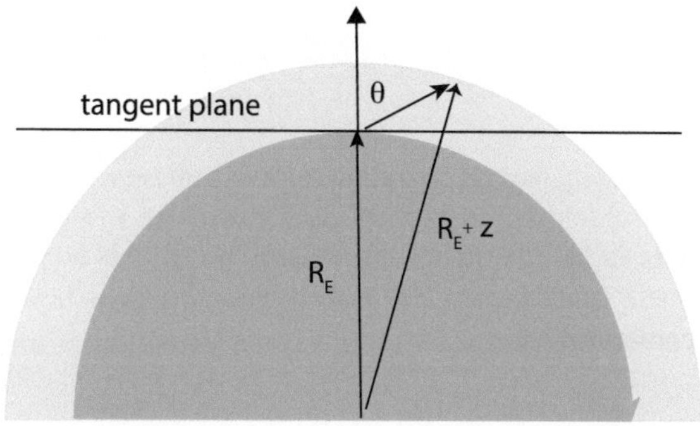

Fig. 3.9 The geometry for the calculation of the atmospheric mass above the tangent plan

Another important factor whether or not the collision can remove the portion of the atmosphere above the tangent plane (see Fig. 3.9) is that the mass of the incoming projectile must exceed the value

$$m* = \frac{2\pi P_0}{g} HR \tag{3.75}$$

where P_0 is the atmospheric pressure, g the acceleration of gravity, H the scale height and R the radius of the planet. It can be shown that the mass above the tangent plane is given by

$$m_{tp} = M_{atm}\frac{H}{R}, \qquad H << R \tag{3.76}$$

For Mars $m_+ \approx 4 \times 10^{13}$ kg corresponding to a silicate object of ≈3 km diameter. An interesting consideration can be made if we consider a swarm of objects. In this case after the first impact the atmospheric mass is reduced so it is pressure and further erosion can occur more easily. If we indicate the mass flux as $M_{cum}(m*)$ then the atmospheric mass loss is given by

$$\frac{dM_{atm}}{dt} = M_{cum}(m*) \times 4\pi R^2 \times m_{tp} \tag{3.77}$$

Where $M_{cum}(m*)$ can be parameterized as $\approx am*^{-b}$. Considering that m_{tp} is given by (3.76) and $m*$ by (3.75) were pressure $P_0 = M_{atm}g/4\pi R^2$ we obtain a simple differential equation for M_{atm}

$$M_{atm}^{b-1} dM_{atm} = -2\pi a(RH)^{1-b} \tag{3.78}$$

With the minus sign meaning that the mass is decreasing. Integrating (3.78) we have the solution

$$\frac{M_{atm}(t)}{M_0} = \frac{P(t)}{P_0} = \left(1 - \frac{t}{t*}\right)^{1/b} \tag{3.79}$$

with

$$t^* = \frac{1}{2\pi ab(RH)^{1-b}} \left(\frac{2\pi P_0}{g}\right)^b \tag{3.80}$$

The present parameterization refer to a phase which follows the heavy bombardment phase. Melosh and Vickery have calculated $t \approx 60/\text{Gyr}$ that is 60 billion years for Mars and even longer for the Earth. So there is no possibility that the present atmosphere could be eroded. However the previous heavy bombardment phase can be represented by

$$N_{cum}(m,t) = a(1 + Be^{-\lambda(t+4.63)})m^{-b} \tag{3.81}$$

Where $B = 2{,}300$ and $\lambda = 4.63\text{Gyr}^{-1}$. In this case the solution is the following

$$\frac{M_{atm}(t)}{M_0} = \frac{P(t)}{P_0} = \left[1 - \frac{t}{t^*}\frac{Be^{-4.6\Lambda}}{\lambda t^*}(1 - e^{-\lambda t})\right]^{1-b} \tag{3.82}$$

This expression gives an initial atmospheric pressure on Mars of about 1 bar. The conclusion is that impact erosion was significant in the early stage of the solar system.

References

Catling, D.C., Kssting, J.F.: Atmospheric Evolution on Inhabitated and Lifeless Worlds. Cambridge University Press (2017)

Fegley, B., Ladders, K.: Chemistry of the Solar System, Royal Society (2010)

Holland, H.D.: The chemical Evolution of the Atmosphere and the Oceans. Princeton University Press (1984)

Horedt, G.P.: Blow off of the protoplanetary cloud by a T-Tauri like solar wind. Astron. Astrophys. **64**, 173 (1978)

Hunten, D.M., Pepin, R.O., Walker, C.G.: Mass Fraction. Hydrodyn. Escape Icarus **69**, 532 (1987)

Hunten, D.M.: The escapeof light gases from pkanetary arÃtmospheres. J. Atmos. Sci. **30**, 481 (1971)

Hunten, D.M.: Thermal and nonthermal escape mechamisms for terrestrial bodies. Planet. Space Sci. **30**, 773 (1982)

Kasting, J.F.: Earth's early atmosphere. Science **259**, 920 (1993)

Lewis, J.S., Prinn, R.G.: Planets and their Atmospheres: Origin and Evolution. Academic Press (1987)

Lewis, J.S.: Physics and Chemistry of the Solar System. Academic Press (2010)

Pepin, R.D.: On the origin and early evolution of terrestrial planet atmospheres and meteoritic volatiles. Icarus **92**, 2 (1991)

Sekiya, M., Nakaza, K., Hayashi, C.: Dissipation of the primordial terrestrial atmosphere due to irradiation of the solar EUV. Progr. Theor. Phys. **61**, 1980 (1968)

Vickery, A.M., Melosh, H.J.: Atmospheric erosion and impactor retention in large impacts, with application to mass extinctions. In: Global Catastrophes in Earth History; An Interdisciplinary Conference on Impacts, Volcanism, and Mass Mortality: Geological Society of America Special Paper 247 (1990)

Walker, C.G.: Evolution OI the Atmo Sphere. McMillan (1977)

Walker, C.G.: Impact Erosion of planetary atmospheres. Icarus **68**, 87 (1986)

Zahnle, K.J., Kasting, J.F.: Mass fractionation during transonic escape and implications for loss of Water from Mars and Venus. Icarus **68**, 462 (1986)
Zahnle, K.J., et al.: Emergence of a Habitable planet. Space Sci. Rev. **129**, 35 (2007)

Chapter 4
The Ancient Atmosphere

In the previous chapters we have learned the basics on the origin of the atmosphere of the Earth. The main conclusion is that the atmosphere was not captured from the nebular gas but was released from the impacts of planetesimals with the Earth surface. There were huge amounts of water vapor that made up the primitive atmosphere. The subsequent cooling of the planet caused the water vapor to condense and forms the oceans. A good part of the remaining water in the atmosphere was split by photodissociation with the hydrogen and other light gases escaping. One possible proof of this process is the abundance of the inert gases. The recent formed mantle would release volcanic gases starting about 4.0 billion of years ago that would produce a reducing atmosphere which was made up of hydrogen, methane and carbon dioxide that would provide the environment for the origin of life.

4.1 The Reduced Early Atmosphere

Theories on the origin of life on Earth requires a reducing atmosphere composed then by H_2 (molecular hydrogen), CH_4 (methane) and NH_3 (ammonia). Initially it was thought that such an atmosphere could be provided by volcanic gases whose composition has not changed since the formation of Earth' core about 30 Ma after the planet was completed. This episode subtracted most of the iron from the mantle so that the volcanic emission would be composed of non reduced gases like H_2O (80–90%), CO_2 (6–12%), $CO_2 \approx$ (3%) the rest being constituted by SO_2 and H_2S. This composition does not include significant quantities of reducing gases so that to justify the presence of a reducing atmosphere theories were developed that attributed the release of such gases to impacts from the remains of planetesimals. Recently, Kevin Zahnle and coworkers have outlined a possible scenario were the *late veneer* scenario contained the collision which generated the Moon. Following an earlier paper by Norman Sleep and the same Kevin Zahnle we assume that our planet collided

G. Visconti, *Origin and Evolution of the Atmosphere and Ocean*,
https://doi.org/10.1007/978-3-032-02653-8_4

during the first 100 Myr after its completion with a Mars sized object (≈3000 km in diameter). The impact melted part of the existing mantle and released gases only partially soluble in the melt like H_2O and carbon dioxide that constituted a thick atmosphere. After the water condensed into the ocean a 100 bar CO_2 atmosphere survived. We now refer mainly to the Zahnle paper to introduce the concept of *quenching*.

4.1.1 Quenching

The concept of chemical quenching in the atmospheres of giant planets was introduced by Prinn and Barshay in 1977 in a *Science* paper. They were studying the early results for the presence of carbon monoxide in the Jupiter atmosphere and assumed that CO was transported from the lower hot atmosphere of the planet and converted to methane (CH_4) by the reaction

$$H_2 + CH_2O \rightarrow OH + CH_3$$

So that they introduced a lifetime for CO given by

$$t_{chem} = \frac{[CO]}{k[H_2][CH_2O]}$$

With k reaction constant. They compared this lifetime with a convective one defined as

$$t_{conv} = \frac{H^2}{K}$$

where H is the scale height and K the diffusion coefficient. Whenever $t_{chem} \approx t_{conv}$ in essence, vertical mixing results in CO in an air parcel being *chemically quenched*, The same concept was repeated in the Lupu et al. (2014) paper. They claim that *sufficiently rapid vertical mixing rates relative to chemical reaction rates cause departures from chemical equilibrium in the atmospheres of gas giant planets*. That is at low altitude with temperatures high enough chemical reactions proceed rapidly and slow down with increasing altitude with lower temperature and pressure until $t_{chem} \approx t_{conv}$. Above this level chemical equilibrium is not attained and the gases are quenched. That is CO is being replaced as fast as it is consumed. However in the work of Zahnle et al. (2020) the concept of quenching does not invoke transport but simply refer to a slow down of chemical reactions until they stop. The author in the paper say *As the gas cools, chemical reactions between the more stable molecules slow down until for all practical purposes they stop and the gas composition is said to have quenched or frozen* (Zel'dovich Ya and Raizer Yu 1967). Actually in this voluminous book this term is mentioned only once. In this respect the concept was enunciated by Chameides and Walker in a 1986 paper which deal with thermochemistry in presence of lightning. They affirm *As the air cools, mainly by expansion, the rates of reaction decrease until they become too slow to maintain the concentration of a given molecule at the equilibrium value appropriate to the temperature. The temperature at which the concentration of a given species departs from equilibrium is called the 'freeze out' temperature for that constituent because the mixing ratio corresponding to equilibrium at that temperature becomes frozen into the gas.*

The importance of quenching arises because it is invoked by the latest theories on the origin of reducing atmospheres (Zahnle et al. 2020; Itcovitz et al. 2022). The most important parameter is the temperature, a shock heated atmosphere would quench to CO, N_2 and H_2.

4.2 The Impact Generated Atmosphere

The calculation involves the collision of a Mars sized object with the Earth. This has a mass very similar to the present one (5.9×10^{24} kg) and has a surface ocean containing roughly 1.85 equivalent ocean which amount to 2.53×10^{21} kg of water. This data divided by the surface of the Earth (5.15×10^{14} m^2) would give an equivalent of pressure of 500 bar. The atmosphere would contain 100 bar of CO_2 and 2 bar of nitrogen N_2. The impact would melt part of the existing mantle and vaporize most of the water then present and forms molecular hydrogen through the reaction

$$Fe + H_2O \rightarrow FeO + H_2$$

Two other important reduced components like methane (CH_4) and ammonia (NH_3) would be produced by thermochemical reactions. We can make some energy consideration. To boil 1 kg of ocean water require $2,5 \times 10^6$ J at 1 bar. So to evaporate the entire ocean will require $1.37 \times 10^{21} \times 2.5 \times 10^6 \approx 3.4 \times 10^{27}$ J. A Mars sized object (6.9×10^{23} kg) hitting at 11 kms^{-1} would have enough energy to vaporize the whole ocean. The atmosphere eventually cools sufficiently to the point that the steam rains out from the atmosphere and reforms oceans on the planet surface. The remaining atmosphere is reducing in nature. Itcovitz et al. describe in detail the collision process which results in the formation of the Moon from the debris, erosion of the atmosphere and melting of large part of the mantle. The consequences of these impacts have been treated differently by different authors which make a distinction between large impacts (like the one we just mentioned) and continuous bombardment known as *late veneer*. We refer mainly to the work of Zanhle and coworker which have build a thermochemical model which considers a five gas system including hydrogen (H_2, H_2O, CH_4), Carbon (CO, CO_2, CH_4), Oxygen (CO, CO_2, H_2O) an two chemical reactions that involve the creation of CO_2 and methane

$$CO + H_2O \leftrightarrow CO_2 + H_2 \tag{4.1}$$

And

$$CO + 3H_2 \leftrightarrow CH_4 + H_2O \tag{4.2}$$

Oxygen however could be controlled by a mineral buffer and in this case the following reactions must be considered

$$2H_2O \leftrightarrow O_2 + 2H_2 \tag{4.3}$$

Notice the double arrow that means that such relations should implies to reach thermochemical equilibrium, In this system nitrogen is treated as a minor specie and the two corresponding equilibria are

$$N_2 + 3H_2 \leftrightarrow NH_3 \tag{4.4}$$

While the other involves hydrogen cyanide (HCN)

$$H_2O + HCN \leftrightarrow CO + NH_3 \tag{4.5}$$

These reactions are treated like hydrolysis so we introduce an equilibrium constant for the different reactions

$$k_{41} - \frac{p^2_{NH_3}}{p_{N_2} p^3_{H_2}} \qquad k_{42} = \frac{p_{H_2O} p_{HCN}}{p_{CO} p_{NH_3}} \tag{4.6}$$

where p_i are the partial pressures. In this way the conservation law for nitrogen

$$N_N = N_{NH_3} + N_{HCN} + 2N_{N_2} \tag{4.7}$$

For example combining Eqs. (4.4), (4.5) it is possible to obtain NH_3 (see appendix). At this point it is possible to determine the quenching conditions for the system $H_2 - H_2O - CH_4 - CO - CO_2$ and the other for $H_2 - N_2 - NH_3 - HCN$. For CH_4 we have a low pressure quenching temperature and a high pressure one

$$T_q(CH_4) = \frac{2.5 \times 10^4}{\ln(0.025 t_c p^2)}, \qquad T_q(CH_4) = \frac{4,2 \times 10^4}{\ln(3.3 \times 10^5 t_c p)} \tag{4.8}$$

where t_c is the cooling time and p pressure. Similar expression are found for ammonia

$$T_q(NH_3) = \frac{5.2 \times 10^4}{\ln(1 \times 10^7 t_c p)} \tag{4.9}$$

Numerical experiments simulating the impact of a Vest size object (mass 2.5×10^{20} kg) with an atmosphere containing 2 bar of CO_2, 1 bar of N_2 and 500 bar of water as an ocean result in a quench point for methane of about 750 K at a pressure of 0.1 bar while for ammonia is about 1100 K and $\approx 2 \times 10^{-3}$ bar. As we see the cooling time is essential in these calculations and when the temperature is well below the quenching value the chemistry switches from thermochemistry to *photochemistry*

4.2.1 Cooling Times

First of all we consider the energy necessary to heat an atmosphere made up of carbon dioxide, nitrogen and water vapor and to evaporate the ocean. Such energy is

$$E_{atm} = (Q_w + C_{H_2O}\Delta T)M_{H_2O} + C_{CO_2}\Delta T M_{CO_2} + C_{N_2}\Delta T M_{N_2} \tag{4.10}$$

where C_i are the heat capacities of the gases and M_i their masses and Q_w the latent heat of vaporization. it is interesting to evaluate the first term which correspond to heat the water by 1500 K and vaporize the entire ocean. We have

$$(Q_w + C_{H_2O}\Delta T)M_{H_2O} = (2.5 \times 10^6 + 2 \times 10^3 \times 1.5 \times 10^3) \times 1.4 \times 10^{21} = 7.7 \times 10^{27}\, joule$$

This energy could be compared with the kinetic energy of an object the size of Vesta (2.5×10^{20} kg) hitting the Earth at 17 km/s, In this case

$$\frac{1}{2}mv^2 = 1,25 \times 10^{20} \times 17^2 \times 10^6 \approx 3.6 \times 10^{28}\, joule$$

If only half of this energy could be available it could heat and evaporate 2.3 oceans of water. To evaluate the cooling time we just divide the E_{atm} by the infrared flux of the Earth. In this case the temperature difference is between 1400 K and the quench value. The emitted infrared flux F_r is taken $\approx$150 wm^{-2} which correspond to an albedo 0.5 and a Sun about 85% dimmer than today. The cooling time for methane is then

$$\begin{aligned} t_c(CH_4) &= \frac{C_{H_2O}M_{H_2O} + C_{CO_2}M_{CO_2} + C_{N_2}M_{N_2}}{4\pi R^2 F_{ir}}(T - T_c) \\ &= \frac{2 \times 10^3 \times 1.4 \times 10^{21} + 800 \times 5.14 \times 10^{21} + 1100 \times 5.14 \times 10^{19}}{150 \times 5.14 \times 10^{14}}(1400 - 800) \\ &\approx 1700\ years \end{aligned} \tag{4.11}$$

For ammonia due to the fact that the quench temperature is higher we obtain a lower cooling time. In the appendix we show a much detailed treatment of this topic.

4.3 Chemistry of Reducing Atmosphere

Until a few years ago it was thought that the reducing atmosphere was essentially produced by gases emitted by the volcanoes or the magma ocean, These gases were mainly CO, CO_2 and sulfur bearing gases. The atmosphere was quite thick (hundreds of bar?) and contained the water vapor as residual of the accretion process. In the last decade a new picture has emerged which attribute the origin of reducing atmosphere to a giant impact that could have coincided with the collision that generated the Moon followed by a stream of minor impacts known as late veneer. Kevin Zahnle ad coworkers have in mind a scenario which has been illustrated in their paper (Zahnle 2006). After the collision the temperature of the planet could have reached 3000 K maintaining a molten magma ocean. At this point there existed a magma ocean and an atmosphere made up of water vapor and carbon dioxide ($\approx$100 bar). The magma ocean was maintained by the greenhouse effect produced by the massive atmosphere for about 2 million year. When the mantle solidified the geothermal heath could not support the steam atmosphere which condensed to form a warm ($\approx 500K$) ocean. The cooling of the planet was related to the burial of CO_2 that reached 1 bar pressure after about 20 Myr. We can then distinguish between two atmospheric stages the first was dominated by thermochemistry and the last one by photochemistry when the atmosphere cooled down. The first phase could be influenced whether the surface of the planet was Continental Crust (CC) or Bulk Silicate Earth (BSE). For reference, the BSE is the composition of the silicate portion of the Earth that evolved into the present-day mantle, crust, oceans, and atmosphere. The mantle comprises 99.4% by mass of the BSE. The CC would have degassed mainly silica-rich (felsic rocks) while BSE would have produced *mafic rocks* containing MgO and FeO. Lupu and her coworkers have build a model for this primitive atmosphere whose results can be summarized as follows for the BSE case. A surface pressure of 100 bars and a surface temperature of 1600 K would implies a surface density of $\approx$30kg/m^3 with a composition mainly of CO_2 and H_2O and we see that the temperature drops from 1600 to 1100 K in about 90 bars, This implies a temperature lapse rate of about

$$\frac{dT}{dz} = -\rho g \frac{dT}{dp} \approx 30 \times 9.8 \times 5.10^{-4}/9 \approx 16 \times 10^{-3}\ \mathrm{K\ m^{-1}}$$

Such large lapse rate would extend the troposphere up to about 150 km where the pressure would reduce to about 10 bars. We see the effect of introducing an eddy diffusion coefficient which has the effect of mixing some component like H_2O and CO_2 and quench some other like NH_3 and CH_4 which have quench temperature in the range 800–1000 K. The very deep troposphere would contain the major gases (carbon dioxide, water vapor) and with lower mixing ratio (but again with constant values) nitrogen, sulfur dioxide and hydrochloric acid. All the other gases would occupies the layer above the troposphere. We will talk about the climate in a few pages and for the moment we can assume that about 3.8 billion years ago the atmosphere of the Earth was composed mainly of nitrogen, hydrogen and methane. In this environment it was important the *redox budget* that is the balance between the emitted hydrogen from volcanoes and that escaping to space. At the present time this balance is based on the inventory of oxygen. For example an oxidized compound like carbon dioxide is reduced in the photosynthesis reaction

$$CO_2 + H_2O \xrightarrow{\text{light}} CH_2O + O_2 \qquad (4.12)$$

When the organic material CH_2O is buried for each mol of this compound one mol of O_2 is liberated. In the early Earth the oxygen mixing ratio is so low that it is convenient to use a reduced species for the budget that is H_2 for example which is related to oxygen by the reaction

$$2H_2 + O_2 \rightarrow 2H_2O \qquad (4.13)$$

We may distinguish between an atmospheric budget and a global one which involves ocean and crust. The source of hydrogen must be then balanced by escape flux which is estimated as $0.5 - 2.4 \times 10^{12}$ mol $year^{-1}$. The total production of O_2 should have been according to (4.13) around 1.2×10^{12} mol $year^{-1}$. Now the rate of production of oxygen in the *prebiotic atmosphere* could attributed for example to the reactions

$$\begin{aligned} CO_2 + h\nu &\rightarrow CO + O \\ H_2O + h\nu &\rightarrow OH + H \\ O + OH &\rightarrow O_2 + H \end{aligned} \qquad (4.14)$$

These reactions may not be a net source of oxygen because both CO and H_2 may recombine and form back carbon dioxide and water. Finally however the escape flux of H_2 as we have seen has to be dependent on the total mixing ratio of hydrogen $f_T(H)$ and some limiting flux. It results (see appendix)

$$\Phi_l(H) \approx 2.5 \times 10^{13} f_T(H) \text{atoms cm}^{-2}\text{s}^{-1} \qquad (4.15)$$

The total mixing ratio at the homopause is the sum of the single mixing ratios

$$f_T(H) = f_H + 2f_{H_2} + 2f_{H_2O} + 4f_{CH_4} \qquad (4.16)$$

In the primitive atmosphere this sum is dominated by the water term which could be assumed around 3 ppm (3×10^{-6}) so that the escape flux could be

$$\Phi_l(H) \approx 2.5 \times 10^{13} \times 3 \times 10^{-6} \approx 7.5 \times 10^7 \text{ atoms cm}^{-2}\text{s}^{-1} \qquad (4.17)$$

This number can be converted in mole per year of oxygen

$$4\pi R^2 \times 3.75 \times 10^7 \approx 1.9 \times 10^{26} \; molecules \; O_2 \; s^{-1} \approx 1 \times 10^{10} \text{mol yr}^{-1} \qquad (4.18)$$

This is about 120 less the value previously estimated (4.13). Notice that the production of O_2 is half the escape rate of H_2. However we may assume that hydrogen was escaping at the limiting flux rate so that we should have

$$2.5 \times 10^{13} f(H_2) \approx 2 \times 10^{10} \qquad (4.19)$$

Notice we have used the mixing ratio of molecular hydrogen and we get

$$f(H_2) \approx \frac{2 \times 10^{10}}{2.5 \times 10^{13}} = 8 \times 10^{-4} \tag{4.20}$$

That is 800 ppm. At the present time the H_2 mixing ratio is about 0.55 ppm so that in the prebiotic atmosphere the mixing ratio was about 1500 times higher. The reason for that could be a much higher volcanic degassing rate, We have seen however larger mixing ratios produced by impacts. The numbers we have given could change but the basic conclusion remain that hydrogen mixing ratio was much higher. In the next section we can now summarize what we know about the chemical composition of the primitive atmosphere.

Oxygen, O

In the archean (between 2.5 and 4 billion year ago) the oxygen was produced abiotically because of the absence of photosynthesis. This was done through reactions (4.14) where the atomic oxygen produced by the photodissociation of carbon dioxide was converted in molecular oxygen by

$$O + O + M \rightarrow O_2 + M \tag{4.21}$$

This simple reaction scheme could be inserted in a photochemical model and would produce for example one dimensional profiles like those shown in Fig. 7 in the paper by Kasting and Pollack (1984). The resulting profile for oxygen shows an increase in mixing ratio with the increase of the source (carbon dioxide). At 40 km we observe an increase from 10^{-17} to 10^{-11} gig from 1PAL to 1000 PAL of carbon dioxide. These results refer to a H_2 flux of 4×10^9 cm^{-2}s^{-1} and 1 bar of CO_2.

Carbon Dioxide, CO_2

As we have seen before that CO_2 could have reached 1 bar pressure after the Moon forming impact. However its formation again was a product of photochemistry because most of the carbon was released as CO that was converted in CO_2 according to the following scheme

$$\begin{aligned} H_2O + h\nu &\rightarrow OH + H \\ CO + OH &\rightarrow CO_2 * H \\ H + H + M &\rightarrow H_2 + M \\ net : CO + H_2O &\rightarrow CO_2 + H_2 \end{aligned} \tag{4.22}$$

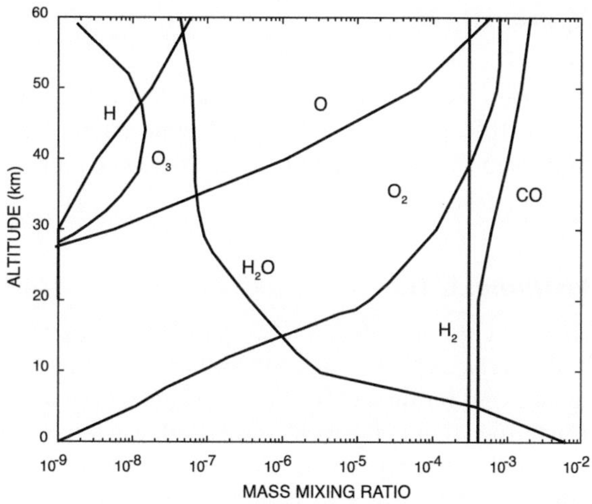

Fig. 4.1 A possible composition of Archean atmosphere

We will examine in detail this topic later on but for the moment we may observe that the fate of such carbon is that of being processed in the crust by the Urey reaction/indexUrey!reaction

$$CO_2 + CaSiO_3 \leftrightarrow SiO_2 + CaCO_3 \tag{4.23}$$

Weathering of rocks on land dissolves carbonates and releases Ca_2^+ from silicates. Calcium ions could react with water and reverse reaction (4.23) in carbonate weathering

$$H_2O + CO_2 + CaCO_3 \leftrightarrow Ca^{++} + 2HCO_3^-. \tag{4.24}$$

The net effect of this process is to remove CO_2 from the atmosphere and acts also as temperature buffer. The presence of a glaciation around the end of the Archean (2.2-2.4 Ga) would indicate a CO_2 pressure of about 0.2 bar for reason related to the greenhouse efficiency at a time when the Sun had a lower luminosity than today(see the *the faint sun paradox*. In more recent epoch when the first geologic records become available the carbon dioxide partial pressure would reach 3 PAL (Present Atmospheric Level) or 0.001 bar.

Nitrogen, N_2

Nitrogen was mainly released during impacts and the pressure during the Archean would be around 2–3 times than today. The excess N_2 would be incorporated into the mantle. The large presence of nitrogen rising the pressure would broaden the CO_2 absorption lines and contribute to the greenhouse effect (Fig. 4.1).

Hydrogen, H_2

We have seen that also for large impact the H_2 mixing ratio would reach values around 1000 ppm only in the upper atmosphere. The most recent results show that early estimate of mixing ratio up to 10% were unrealistic.

Methane, CH_4

In the early Earth a source for methane could be the mineral serpentine. This has an idealized formula $(Mg, Fe)_6Si_4O_{10}(OH)_8$ and some brucite $(Mg, Fe)(OH)_2$. The ferrous iron in the rocks reacts with water to form hydrogen gas and magnetite Fe_3O_4. The hydrogen with CO_2 provides a substrate for modern methanogens at serpentine-fed vents. Serpentine also forms by reaction of water with olivine in basalt and komatiite. The schematic reaction could be

$$CO_2 + 4H_2 \rightarrow CH_4 + 2H_2O \tag{4.25}$$

Or produce directly organic matter (CH_2O)

$$CO_2 + 4H_2O \rightarrow 2CH_2O + 2H_2O \tag{4.26}$$

Methanogen bacteria would appear around 3.5 Ga and we will talk about this later. However various evidence shows that the surface temperature in the Archean was maintained between 0 and 40° by a feedback mechanism involving methane and carbon dioxide (see chapter on climate) that could be summarized by the reaction

$$XSiO_3 * CO_2 + H_2O \leftrightarrow XCO_3 + SiO_2 + H_2O \tag{4.27}$$

where X is a cation. When the partial pressure of CO_2 is high then pH is low and this reaction lower the carbon dioxide mixing ratio and the temperature. Mixing ratio of methane could have reached 1000 ppmv much larger that today 1.8 ppmv.

Carbon Monoxide, CO

Carbon monoxide could be produced by large impacts. However photochemical models have been produced which gives results like those shown in Fig. 4.1 where an atmosphere very similar to the present one is used except for the much higher mixing ratio (0.2 bar) for CO_2. This is responsible for the production of CO

$$CO_2 + h\nu \rightarrow CO + O \tag{4.28}$$

Table 4.1 Summary of major components of the Hadean atmosphere

Component	Quantity	Epoch before present (Ga)
O_2	$0.1 - 3 \times 10^{-5}$	2.4
O_3	$\approx 10^{15}$ molecules m^{-2}	> 2.4
N_2	1-1.1 bar	3-3.5
HCN	$\approx$ 100 ppmv	> 2.3
N_2O	few ppbv	2.4
CO_2	0.005–0.002 bar	3.5–2.3
CH_4	5000–20 ppmv	3.5–2.3
CO	few ppmv<300 ppmv	Following life
H_2	10–100 ppmv	3.4–2.3

The hydrolysis of CO_2 could have also contributed to production of organic compounds. The mixing ratio for CO near the ground reaches 1.5×10^{-5}. Table 4.1 adapted from Catling and Zahnle gives a summary of the atmospheric composition of the Archean atmosphere.

Appendix

Ammonia

Consider Eqs. (4.6) and (4.7) and remember that number densities are proportional to pressures so we can obtain N_{HCN} and N_{N_2} from (4.6) ans substitute to (4.7)

$$N_N = NH_3 + \frac{K_{42} N_{CO} N_{NH_3}}{N_{H_2O}} + \frac{N^2_{NH_3}}{k_{41} N_{H_2}} \tag{4.29}$$

we obtain a second degree equation in N_{NH_3}

$$\frac{N^2_{NH_3}}{k_{41} N_{H_2}} + N_{NH_3} \left(1 + \frac{K_{42} N_{CO}}{N_{H_2O}}\right) - N_N = 0 \tag{4.30}$$

When N_{NH_3} can be calculated once the constant and concentrations are known (Fig. 4.2).

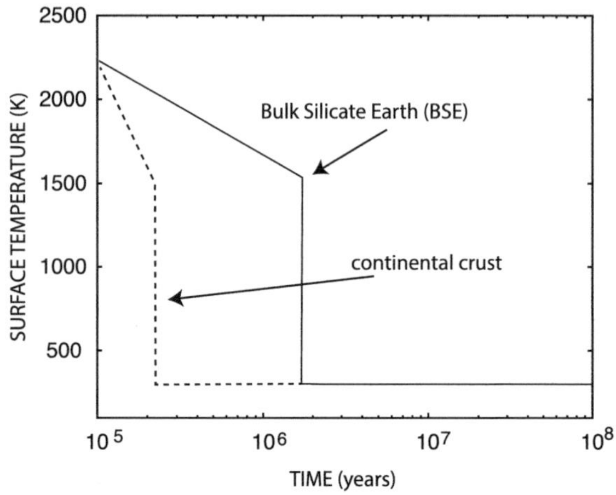

Fig. 4.2 The qualitative thermal history of the Earth after a major impact

Cooling Times

We already mentioned the planet cooling in previous chapters but now we want to include the blanketing effect of the atmosphere. We characterize the planet with a magma ocean (produced by a large impact) over a solid mass with a surface temperature T_{surf} while the particles of the interior have potential temperature T_i. That would be the parcel temperature when brought adiabatically at the surface, Define different temperatures. T_{liq}, liquidus above which the mantle is fully molten, solidus T_{sol} below which the mantle is fully solid and an intermediate temperature critical T_{crit}, which define the molten fraction ϕ

$$\phi = \begin{cases} 1, & T_i > T_{liq} \\ (T_i - T_{sol})/(_{liq} - T_{sol}) & T_{sol} < T_i < T_{liq} \\ 0, & T_i < T_{sol} \end{cases}$$

Now we need the heath capacity of the Earth after the giant impact as the sum of those for the mantle and the core

$$C_v M_\oplus = C_v M_{man} + C_c M_{cor} = 6.2 \times 10^{30} J \text{kg}^{-1} K^{-1} \tag{4.31}$$

where $C_v = 1.2 \times 10^3 J\text{kg}^{-1}K^{-1}$ is the heat capacity of the mantle and $C_c = 650 J\text{kg}^{-1}K^{-1}$ is that for the core. While mantle freeze the effective capacity is

$$C' M_\oplus = C M_\oplus + \frac{Q_m M_{man}}{T_{liq} - T_{sol}} = 1 \times 10^{30} j K^{-1} \tag{4.32}$$

where $Q_m = 4 \times 10^5 jkg^{-1}$ is the heat of fusion of silicates and $M_{man} = 4.1 \times 10^{24}$kg is the mass of the mantle. As we have seen before the heat flow F is written as

$$F = C_0 \frac{k_c(T_i - T_{surf})}{L} Ra^n \tag{4.33}$$

In our case we take $T_{liq} = 1800K$ and $T_{sol} = 1400K$ Where C_0 is a constant 0.09, $k_c = 300 jm^{-2}s^{-1}$ is the thermal conductivity, L is a typical distance (thickness of the mantle) $2.8 \times 10^6 m$, Ra the Rayleigh number and $n = 1/3$. We recall the definition of Rayleigh number

$$Ra = \frac{\alpha g(T_i - T_{surf})}{k\nu} L^3 \tag{4.34}$$

where $\alpha = 2.4^{-5} K^{-1}$ is the expansion coefficient, g the gravity acceleration, ν the kinematic viscosity and $k = k_c \rho / C_v$ he thermal diffusivity. As we have seen the viscosity is a strong function of temperature and the heat flux combining (4.33) and (4.34) can be written as

$$F = C_0 k_c (T_i - T_{surf})^{1+n} L^{3n-1} \left(\frac{\alpha g}{\nu k}\right)^n \tag{4.35}$$

And the cooling of the planet interior is described by

$$\begin{aligned} CM_\oplus \frac{dT_i}{dt} &= -4\pi R_\oplus^2 F, \qquad T_i > T_{liq}, \quad T_i < Tsol \\ C' M_\oplus \frac{dT_i}{dt} &= -4\pi R_\oplus^2 F, \qquad T_{sol} < T_i < T_{liq} \end{aligned} \tag{4.36}$$

However we need to take into account the incoming solar radiation F_{sun} and the heat capacity of the atmosphere C_A, the changes in the temperatures T_i and T_{surf} are

$$\begin{aligned} & CM_\oplus \frac{dT_i}{dt} + C_A \frac{dT_{surf}}{dt} = -4\pi R_\oplus^2 \sigma T_{eff}^4 + \pi R_\oplus^2 F_{sun}(1 - A) \\ & T_i > T_{liq}, \qquad T_i < Tsol \\ & C' M_\oplus \frac{dT_i}{dt} + C_A \frac{dT_{surf}}{dt} = -4\pi R_\oplus^2 \sigma T_{eff}^4 + \pi R_\oplus^2 F_{sun}(1 - A) \\ & T_{sol} < T_i < T_{liq} \end{aligned} \tag{4.37}$$

From our previous result we know that a characteristic time is of the order $\tau \approx CM_\oplus T_i / F \approx 1000 \times 10^{24} \times 2000/4 \times 10^{13} \approx 10 Ga$. For million years after the impact the cooling is linear up to the point when the temperature drops below the T_{liq}. At that point the convection in the mantle increase drastically and the balance switches to the radiative one (Eq. (4.37) with a dramatic drop in temperature.

Photochemistry

We will illustrate an example of simple photochemistry which is based on a very old approach reported by Walker (1978), He starts by calculating the mixing ratio of hydrogen at the ground based on a simple balance between that emitted by the volcanic source ϕ_{vol} and that lost by escape ϕ_{esc} and we made already such evaluation in (4.20). However we need to take into account that part of this flux was lost to recombination with oxygen produced by the photodissociation of water

$$H_2O + h\nu \rightarrow H + OH \tag{4.38}$$

This reaction had a rate J_{wv} (see following paragraph) which is measured in s^{-1} so that total flux of oxygen is simply $\phi_O = J_{wv} N_{H_2O}$ where N_{H_2O} is the columnar density of water vapor in $molecules/\text{cm}^2$. So based on (4.10) the hydrogen mixing ratio is given by

$$f(H_2) = \frac{\phi_{vol} - \phi_O}{2.5 \times 10^{13}} \tag{4.39}$$

This relation gives values of the order of 10 ppmv that roughly agree with those found for a giant impact when transport is absent. To find the oxygen mixing ratio we need to use a more elaborate model with reactions listed below

$$\begin{aligned} H_2O + h\nu &\rightarrow H + OH, & J_{wv} \\ H + O_2 + M &\rightarrow HO_2 + M, & k_2 \\ HO_2 + H &\rightarrow OH + OH, & k_3 \\ OH + H_2 &\rightarrow H_2O + H, & k_4 \\ \text{net}\quad 2H_2 &+ O_2 \rightarrow 2H_2O \end{aligned} \tag{4.40}$$

From these reactions we can write the concentration of oxygen as

$$[O_2] = \frac{J_{wv}[H_2O]}{k_2[H][M]} \tag{4.41}$$

where [M] is third component (for example N_2). However this is very poor approximation because water vapor is fixed by thermodynamic conditions. We will give some numbers after calculating J_{wv} for the moment could be instructive to calculate the concentrations of the radical OH and H. In this case we assume that all the radicals produced by the photodissociation is consumed by k_4 so that

$$[OH] = \frac{J_{wv}}{k_4[H_2]} \tag{4.42}$$

And the sane for [H]

$$[H] = \frac{2J_{wv}}{k_5[M]} \tag{4.43}$$

where we have considered the recombination

$$H + H + M \rightarrow H_2 + M, \qquad K_5 \tag{4.44}$$

Now let us calculate the photodissociation coefficient.

Photodissociation Coefficient

Consider some level in the atmosphere where we have a solar flux in the vertical direction $F(\lambda, z) = F_0(\lambda)\ exp[-\tau(\lambda, z)]$ where τ is the optical thickness. For the sake of computation usually we use the flux $\Delta F(\lambda, z)$ contained in the interval $\Delta\lambda$ and expressed in *photons* $\mathrm{cm}^{-2}\mathrm{s}^{-1}$. The absorbing gas (of which we like to evaluate the photolysis) has an absorbing cross section $\sigma(\lambda)$ measured in cm^2. Then we define photodissociation coefficient the quantity

$$J = \int_{\lambda_1}^{\lambda_2} \Delta F(\lambda, z)\sigma(\lambda)d\lambda \tag{4.45}$$

where (λ_1) and (λ_2) is the interval of interest. When this procedure is applied with the data reported by Walker the results is $J_{H_2O} H_2O = 2 \times 10^{13}\ molecules\ \mathrm{cm}^{-2}s^{-1}$. It is clear that in this case we are talking about columnar density of water vapor. Just for having an idea of the numbers we fix the reactions constants as $k_2 = 6 \times 10^{-32}\ \mathrm{cm}^6\ s^{-1}$, $k_4 = 5 \times 10^{-15}\ \mathrm{cm}^3\ s^{-1}$ and $k_5 = 1 \times 10^{-32}\ \mathrm{cm}^6\ s^{-1}$. These are integrated along the troposphere. Then we adopt $[M] = 2 \times 10^{19}\mathrm{cm}^{-3}$. The resulting values are typically of the order of 1.8×10^{-5} for the $[H_2]$ mixing ratio and number densities of $7 \times 10^9\ \mathrm{cm}^{-3}$ for [H], $5 \times 10^6\ \mathrm{cm}^{-3}$ for [OH] and 10^4 for $[O_2]$. It looks like oxygen as a minor components of the primitive atmosphere. We have to wait a few chapters before find its raise to the present levels.

References

Catling, D.C., Kasting, J.F.: Atmospheric Evolution on Inhabitated and Lifeless Worlds. Cambridge University Press (2017)

Catling, D.C., Claire, M.W.: How Earth's atmosphere evolved to an Oxic state: a status report, Earth and Planet. Sci. Lett. **237**, 1m (2005)

Catling, D.C., Zahnle, K.J.: The Archean atmosphere. Sci. Adv. **6**, 1 (2020)

Chameides, W.L., Walker, J.C.G.: Rates of fixation by lightning of Carbon and Nitrogen in possible primitive atmospheres. Origins Life **11**, 291 (1981)

Fegley, B., Ladders, K.: Chemistry of the solar system. R. Soc. (2010)

Itcovitz, J.P., et al.: Reduced atmospheres of post-impact worlds: the early earth planet. Sci. J. **3**, 115 (2022)

Kasting. J.F., Pollack, J.B.: Effects of High CO_2 levels on surface temperature and atmospheric Oxidation state of the early earth. J. Atmos. Chem. **1**, 403 (1984)

Kasting, J.F.: Bolide impacts and oxidation state of carbon in the Earth's early atmosphere. Orig. Life Evol. Biosph. **20**, 199 (1990)

Lupu, R.E, et al.: The Atmosphere of Earth-like planets after giant impact events. Astrophys. J. **784**, 27 (2014)

Prinn, R.G., Barshay, S-S.: Carbon monoxide on Jupiter and implications for atmospheric convection. Science **198**, 1031 (1977)

Schaefer, L.B.: Fegley, Chemistry of atmospheres formed during accretion of the Earth and other terrestrial planets. Icarus **208**, 438 (2010)

Sleep, N.H., Zahnle, K.J., Lupu, R.E.: 2014 Terrestrial aftermath of the Moon-forming impact. Phil. Trans. R. Soc. Am **372**, 20130172 (2014)

Walker, J.C.G.: Oxygen and hydrogen in the primitive atmosphere. Pageoph **116**, 241 (1978)

Zahnle, K.J., Schaefer, L., Fegley, B.: In: Deamer, D., Szostak, J.W. (eds.), Earth's Earliest Atmospheres, in Perspectives on the Origins of Life. Cold Spring Harbor Press (2010)

Zahnle, K.J.: Earth's earliest atmosphere. Elements **2**, 217 (2006)

Zahnle, K.J., et al.: Creation and evolution of impact-generated reduced atmospheres of early earth. Planet. Sci. J. **1**, 11 (2020)

Zel'dovich, Y.B., Raizer, Y.P.: Physics of Shock Waves and High Temperature Hydrodynamic Phenomena. Academic Press (1967)

Chapter 5
Origin and Evolution of the Oceans

We have seen up to this point how the atmosphere of the Earth originated and we have vaguely mentioned that water was liberated from the impacts of planetesimals and may be comets creating a steam atmosphere that with the cooling of the planet resulted in a liquid mass, the oceans. We need now to know some proof of this theory together with the knowledge of the chemical evolution of the oceans, how they acquired and maintained the salinity. Oceans probably contributed also to the change in the composition of the atmosphere (i.e., the rise of oxygen) and finally to the origin of life.

5.1 Where the Water Came From?

The planet Earth contains something like 1.7×10^{21} kg of water which represent about $2.8 \times 10^{-4} M_{\oplus}$. Precise estimates of the different reservoirs of water are difficult to make but we can safely assume a total water content of $5.0 \times 10^{-4} M_{\oplus}$. Remember that the total water in the oceans is about 1.4×10^{21} kg. As usual there is a debate from where and when this water came about and we have to justify more than twice the amount of water in the ocean. Geologic evidence shows that oceans were present on Earth already 4.40 Ga while we have seen that the solar system birth was probably dated 4.56 Ga and the Earth was completed in about 100 Ma. There are two hypotheses about how water was delivered. The first is that water was available during the formation of the Earth and the other that water was continuously delivered to the planet by collisions with small comets. According to some space observations the Earth has received between 2.2 and 8.5×10^{21} kg of water during its formation. The correct hypothesis probably is that the Earth received its water through the planetesimal collisions with a final veneer contributed by comets. In order to discuss further we can discuss the data reported by Pinti (2005). These data seem to indicate that the most likely source of water for the Earth are carbonaceous chondrites, (CC).

G. Visconti, *Origin and Evolution of the Atmosphere and Ocean*,
https://doi.org/10.1007/978-3-032-02653-8_5

Considering that their D/H ratio is quite similar to that of the whole Earth. On the other hand, the same data shows that contribution from comets or material captured from the solar nebula (PSN) is less likely. CC contains up 22% in weight of water and hydrous CC originate on the outer region of the asteroid belt between the orbit of Mars and Jupiter. At this point we can formulate a scenario for the origin of the oceans considering also some evidence from sparse geological records. A possible proof of the existence of the oceans is the presence of sedimentary rocks the oldest one date about 3.85 Ga and so then oceans were probably formed just after the planet was completed. Very important for the studies of these early phases are the mineral *zircon* $ZrSiO_4$. Zircon is a common U-rich trace mineral in granitic rocks that preserves a detailed record of the magma genesis. As we have mentioned earlier the Earth collided with a Mars-sized object about 4.5 Ga. This collision (which is probably responsible of the Moon formation) released about 7×10^6 Jkg^{-1} enough to vaporize the silicates on the surface of the Earth ($6 - 14 \times 10^6$ Jkg^{-1}) and raise then temperature up to 2000 °C. This caused the melting of the upper mantle and formed a dense atmosphere of gaseous silicates that cooled down and precipitated in a few thousand years. The residual atmosphere was made mainly of water vapor and its amount was equivalent to the ocean mass; it would correspond to a pressure of $1.4 \times 10^{21} \times 9.8/(4\pi(6.4 \times 10^6)^2) \approx 270$ bars. Another component of the atmosphere was CO_2 whose mass corresponds to what it is found now in the mantle which is comprised between 1.6×10^{20} and 1.1×10^{21} kg which correspond to a pressure up to 40–210 bars. This composition produced a runaway greenhouse effect with the contribution of an internal flux of about 150 wm^{-2}. With the reduction of this flux the runaway could not be maintained and the cooling produced a layer of basaltic composition which separated the atmosphere from the interior. In a few thousands year the temperature was further reduced to permit condensation of water. Considering the high pressure the condensation temperature should have been around 600 K. The oceans formed with heavy precipitation up to a rate of 200 mm/day about 10 times the average of present tropical precipitation. If the atmospheric pressure was of the order of a few hundred bars, the condensation temperature could be around 600 K which is still too hot to think of a possible origin of life. Temperature of the ocean water should have decreased down to 60–100 °C. This level could be reached only when the ocean temperature would be controlled by the atmosphere greenhouse and that means when the internal flux went below 100 w m^{-2}. That could happen by reducing the CO_2 partial pressure between 5–25 bars. That would require to get rid of 205–185 bars of the gas. The most popular suggestion is that of carbonation according to the reactions

$$
\begin{aligned}
CaSiO_3 + H_2O &\rightarrow Ca^{2+} + 2HCO_3^- + SiO_2 \\
Ca^{2+} + 2HCO_3^- &\rightarrow CaCO_3 + H_2CO_3
\end{aligned}
\tag{5.1}
$$

the carbonates produced by the second of (5.1) could be removed by plate tectonics. If we assume that the recycling rate of CO_2 was similar to the present one ($\approx 10^{12}$ kg $year^{-1}$) a CO_2 quantity equivalent to a pressure of 210 bars ($\approx 10^{22}$ kg)

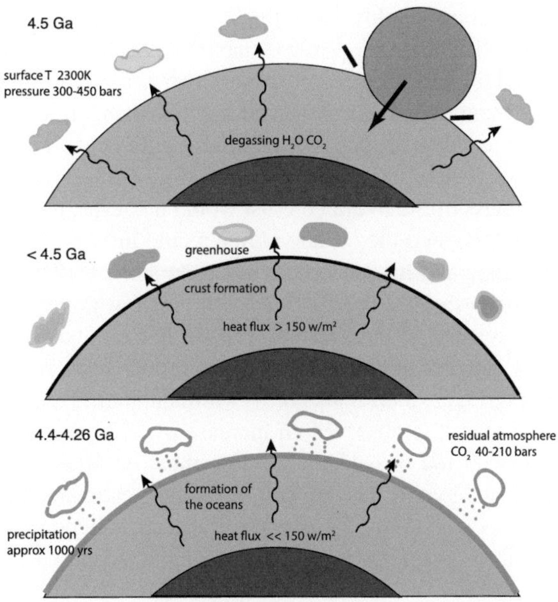

Fig. 5.1 Cartoon showing the ocean formation

could be removed in about 10^9 $year$(1 Ga). We may assume that oceans be stable at the end of the Hadean (3.9 Ga). Figure 5.1 is a cartoon showing the possible chronology of the events. Also in the appendix we report short calculations on the cooling time and some aspect of the runaway greenhouse.

5.2 The Primitive Ocean

The most surprising fact about primitive ocean is the very sparse geological data. We have remarked several times that probably the ocean rained out from a steam atmosphere very early in the Earth history but most of the assumptions are ad hoc. Even the question about the constancy of the volume of the ocean has not been settled. It is then interesting to examine in detail the different characteristics of the primitive ocean.

5.2.1 The Volume of the Ocean

Until recently it was thought that the volume of the ocean has remained constant. The most convincing argument about the constancy of ocean volume is that formulated many years ago by James Kasting and Nils Holms. They begin by observing that at the bottom of the ocean near the ridges the pressure is around 250 bars and the temperature roughly 400 K that is water is near its critical point. This ensures optimal convective heat transport and maximal penetration of hydrothermal circulation along the ridge axes. The oceanic crust is hydrated to a depth of about 1 km and this produces a substantial flux of water to the mantle when this is subducted. This ingassing rate of water should be enough to balance the outgassing rate from the mantle and the process is such to self-stabilize the ocean volume. Should the ocean becomes shallower the thickness of the hydrated crust would decrease, the ingassing rate would be less the outgassing and the ocean volume would increase again. Recently, however, Francis Albarede and his coworkers based on some geological evidence have claimed that the extension of the emerged continents was very much reduced in the Archean with respect to the present. The evidence is based on the phosphorus sediments. This element enters the ocean essentially from runoff while is destroyed in mid-ocean ridges, their flanks or by sedimentation. By examining phosphorus content in mudstone it is observed an increase of 34% per Ga and because there is not inner source in the ocean this increase must be related to the increase in riverine flux produced by the growing continents. It is estimated that in the Archean the emerged land occupied only 7% of the Earth surface in contrast with the present 30%. Later on in this chapter we again talk about the ocean volume.

5.2.2 *Temperature of the Primitive Ocean*

In the case of temperature there is way to get some data from the geologic record. This is the famous *Urey thermometer* which is illustrated in detail in the appendix. It is based on the measurement of the relative abundance of the oxygen isotopes $^{18}O/^{16}O$ expressed by the quantity $\delta^{18}O$. When the ocean water evaporates the vapor is enriched in the lighter isotope (^{16}O) while the ocean water is enriched in the heavier isotope. It can be shown that a positive change in $\delta^{18}O$ in rocks (for example, in cherts) corresponds to a cooling in the water temperature and vice versa. ^{18}O measured on rocks formed in the Archean seem to show that they formed in a seawater that was much hotter than today. Other data seem to indicate a warmer ocean in the Archean as confirmation of most of the model for that period that shows a temperature decreasing from about to 230 °C to the present 20 °C. Another method to study past temperature is through the examination of ancient water inclusions in rocks. The high temperature mentioned is attributed to the runaway greenhouse due to the massive presence of carbon dioxide in the atmosphere. Models of the thermal history of the oceans are available (Tajika and Matsui 1992).

5.2.3 *The Chemical State of the Oceans*

Two factors may have influenced the chemical composition of the primitive ocean. On one side the higher CO_2 pressure in the atmosphere and, on the other side, the fact that erosion could have been different and producing different inputs to the oceans. Very good reviews of the changing composition of the ocean have been published (Hollamd Lowenstein et al. 2014). It is beyond the scope of this book to review the huge amount of data on this subject so that we concentrate on some detail dealing with the pH of the ocean and the role of sulfur compounds. Of particular interest during the Archean (4 to 2.5 Ga) is the absence of gypsum in the deposits of this period. Gypsum is a salt which corresponds to *Calcium Sulfate diHydrate* ($CaSO_4 \cdot 2H_2O$) and its absence could indicate a deficit in sulfate or calcium. This is an example how the evidence from sedimentary records may give indication on the sea composition. Calcium could have been consumed in the formation of calcium carbonate

$$Ca^{2+} + 2HCO_3^- \rightarrow CaCO_3 + CO_2 + H_2O \tag{5.2}$$

In particular calcium is depleted when $[HCO_3] > 2[Ca^{2+}]$ where in this case brackets are indicating the concentration. Many years ago this characteristic suggested to Stephan Kempe and Egon Degens that early ocean could be what they call a *soda lake* where the amount of calcium ions is negligible. These authors report for the largest soda lake known about 150 $meq\ l^{\cdot 1}$ of $HCO_3 + CO_3$ and Cl^- and only 0.4 $meq\ l^{\cdot 1}$ of Ca^{2+}. The unit $meq\ l^{\cdot 1}$ measures the capacity to deliver a mole of H^+ if acid, or OH, if a base. The water in a soda lake has a pH ranging between 9 and 11. However this hypotheses seems to be contradicted by the fact that transition from a bicarbonate ocean to modern one happened between 1.8 and 1.6 Ga and not earlier. A fundamental indicator of the chemical condition of the ancient ocean is the pH that we have calculated in the appendix. As we have shown the concentration of the hydrogen ions is given by

$$[H^+] = \left(K_1 K_2 K_H \frac{P_{CO_2}}{[CO_3^{2-}]}\right)^{1/2} \tag{5.3}$$

where the concentration of carbonate is related to the concentration of dissolved calcium

$$[C_a^{2+}][CO_3^{2-}] = 9 \times 10^{-7} M^2 \tag{5.4}$$

So it is rather intuitive that in ancient times because of the higher carbon dioxide pressure the pH should have been lower the problem being how much lower. The present partial pressure of carbon dioxide is about 0.0004 bars and it can be shown that for a pressure of about 0.3 bars the pH should range between 6.6 and 6.9 so we would have had a more acidic ocean. The higher CO_2 pressure could have been compensated by a lower calcium concentration because of the reduced erosion from the less extension of the continents. The concentrations in (5.3) expressed in *moles/liter* and based on what we show in the appendix it is possible to relate the surface partial pressure of CO_2 to the carbon dissolved in the ocean $CO_2(aq)$

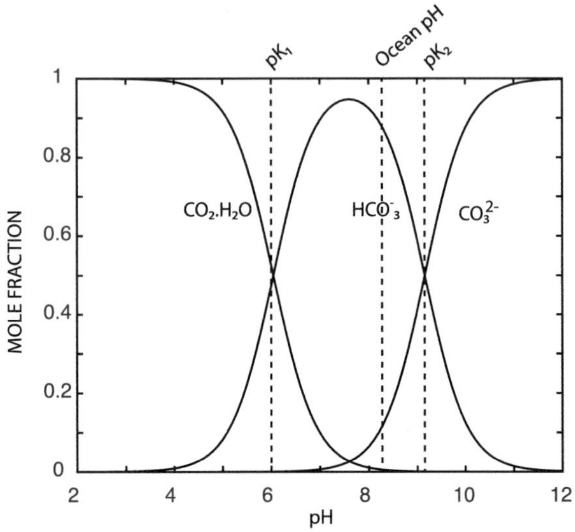

Fig. 5.2 The mole fraction of the three carbon compounds as a function of the pH. Also indicated is the present value of the pH

$$P_{CO_2} = \frac{CO_2(aq)}{K_H}\left(1 + \frac{K_1}{[H^+]} + \frac{K_1K_2}{[H^+]^2}\right)^{-1} \tag{5.5}$$

There are two other relations between the pH and pressure that can be easily obtained

$$\begin{aligned} [H^+][HCO_3^-] &= K_1K_HP_{CO_2} = 10^{-7.6}P_{CO_2} \\ [H^+]^2[CO_3^{2-}] &= K_1K_2K_HP_{CO_2} = 10^{-16.7}P_{CO_2} \end{aligned} \tag{5.6}$$

Equations (5.5)–(5.6) are represented in Fig. 5.2 together with the relation

$$\begin{aligned} [HCO_3^-] = 2[C_a^{2+}] \rightarrow [H^+][HCO_3^-] &= 2[H^+][C_a^{2+}] \\ = 1.8 \times 10^{-6}[H^+])[CO_3^{2-}]^{-1} &= 10^{-7.6}P_{CO_2} \end{aligned} \tag{5.7}$$

The position of the curved line depends on the carbon content and how this is distributed between the solid and fluid Earth. At the present time it is estimated that the carbon present on Earth amounts to $1.8 \times 10^{24}\,g$. Of these about $4 \times 10^{19}\,g$ are in the oceans (Lee et al. 2019) which corresponds to 0.0023 mol/lit. According to the previous considerations this would imply that about 4×10^{15} kg are in the atmosphere generating a pressure of 80 Pa which is the double we observe. It is clear that it is necessary to know the pH of the ocean which increases with the increase of dissolved CO_2. The results of a most recent work by Halevy and Bacham on the evolution of pH show a decrease down to about 6.5 and 7.5 around the GOE. Albarede and coworkers in the mentioned paper set the record straight on the possible chemical composition of the ocean before the rise of oxygen. They argue that because of the reduced extension of the continents the alkalinity flux ($[CaCO_3^-] + 2[CO_3^{2-}]$) was very much reduced which would imply a lower pH and an higher chlorine content.

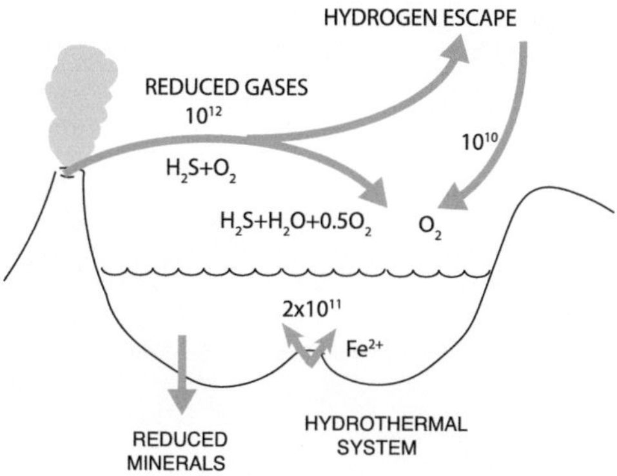

Fig. 5.3 Geochemical cycle of oxygen and reduced species in the early Earth. The fluxes are in mole/year

The lack of oxygen allowed the solubility of iron (Fe) and bring to the formation of methane by the reaction

$$\frac{1}{8}CO_2 + Fe^{2+} + \frac{7}{4}H_2O \Longleftrightarrow \frac{1}{8}CH_4 + FeO(OH) + 2H^+ \qquad (5.8)$$

The lower pH in the ancient ocean together with the low alkalinity flux accounts for the reduced sedimentation of carbonates. One of the main characteristic sediments of the period are *banded iron formation, BIF* whose major contributors at that time were the cations (Na^+, Ca^+, Fe^+, and so on) while now are the carbonates. The charge balance in the present ocean can be written as

$$\begin{aligned} Alk \approx [HCO_3^-] + 2[CO_3^{2-}] \approx [Na^+] + 2[Mg^{2+}] + 2[Ca^{2+}] \\ -[Cl^-] - 2[SO_4^{2-}] \end{aligned} \qquad (5.9)$$

In the Archean ocean could have been quite different

$$[Na^+] + [K^+] + 2[Ca^{2+}] + 2[Fe^{2+}] \approx [Cl^-] + Alk \qquad (5.10)$$

Some of the topics now examined will be discussed again in the coming chapter on biogeochemical cycles.

5.3 Sulfur in the Early Ocean

This paragraph is extracted from our previous book, *The fluid environment of the Earth*. In the pre-biologic ocean (we are talking about 3.8 billion years ago!) sulfate (SO_4^{2-}) was the most abundant sulfur species. It was produced by the oxidation of volcanic sulfur gases. However its concentration was low as compared to modern standards because the inflow of dissolved sulfate from river was low. As shown in

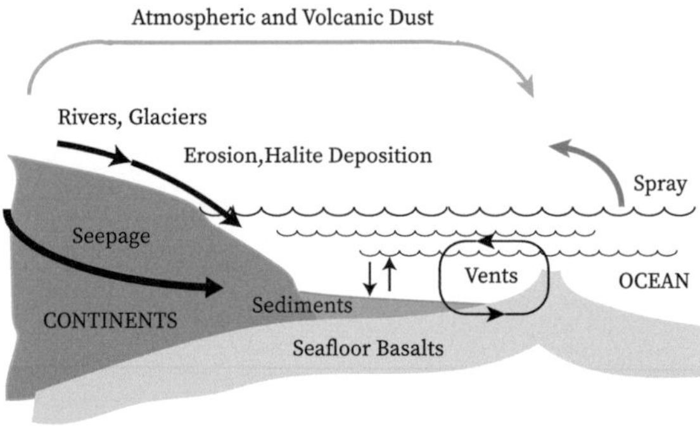

Fig. 5.4 A simple cycle of salt

Fig. 5.4 the sulfur gases like H_2S (hydrogen sulfide) and SO_2(sulfur dioxide) would react with the oxygen present at that time

$$\begin{aligned} H_2S + 2O_2 &\leftrightarrow H_2SO_4, \quad \text{sulfuric acid} \\ SO_2 + H_2O + \tfrac{1}{2}O_2 &\leftrightarrow H_2SO_4 \end{aligned} \tag{5.11}$$

For pressures existing in that atmosphere such reactions would end to the right so that such compounds would be stable. On the other hand, they had very low solubility so that hardly could be rained out and transfer of hydrogen sulfide from the atmosphere to the ocean would have been enhanced by the reaction with dissolved iron. We know for sure that hydrogen sulfide could react with radical [OH]

$$H_2S + OH \rightarrow HS + H_2O \tag{5.12}$$

The radical [HS] (hydrosulfide) would be removed with a lifetime of about one and half day

$$HS + H \rightarrow H_2 + S \tag{5.13}$$

The atomic sulfur could be oxidized back to sulfur dioxide. Possible removal of sulfur atoms could result from the reactions

$$\begin{aligned} S + OH &\rightarrow SO + H \\ SO + OH &\rightarrow SO_2 + H \end{aligned} \tag{5.14}$$

but any sulfur monoxide produced in the atmosphere was more likely to suffer photolysis back to elemental sulfur. Sulfur dioxide would be converted to bisulfite ion

$$SO_2 + OH \rightarrow HSO_3 \tag{5.15}$$

This determines a lifetime of about 3.5 days in the atmosphere for sulfur dioxide. HSO_3 is rather insoluble but it is easily converted into aerosols that could be rained out and provide a source of sulfate to the ocean. When tetravalent sulfur S (IV) enters the composition of sulfite this becomes more soluble and could be oxidized according to

$$4SO_3^{2-} \rightarrow 3SO_4^{2-} + S^{2-} \tag{5.16}$$

In summary, sulfur dioxide emitted by volcano would probably have been photochemically oxidized in the atmosphere to sulfate, which would have been washed into the ocean. Any sulfite or elemental sulfur that escaped oxidation in the atmosphere would probably have been oxidized to the thermodynamically stable S(VI) state by chemical and photochemical reactions in the surface layers of the ocean. The geochemical balance illustrated could be then summarized as all of the sulfate involved in hydrothermal interactions on the early Earth was reduced which involve first the formation of ferrous sulfide (FeS)

$$Fe^{2+} + HS^{-} \rightarrow FeS + H \tag{5.17}$$

And then

$$FeS + H_2S \rightarrow FeS_2 + H_2 \tag{5.18}$$

With pyrite (FeS_2) that would then be precipitated. Sulfate, on the other hand, was almost completely removed by the reaction

$$SO_4^{2-} + H_2S + 2O_2 \tag{5.19}$$

In short if iron (in the +2 oxidation state) is sufficiently abundant relative to sulfate to reduce all sulfate to sulfide, the sulfide will precipitate as iron sulfide minerals, resulting in virtually complete removal of the sulfur from the of circulating seawater. It is estimated a concentration of $10^{-3} mol\, l^{-1}$ while today is about 30 time that value. Recent evaluation gives concentration much lower. However the presence of sulfate in the ocean could be important because it could promote a process of photosynthesis favored by the fact that both iron and sulfur have multiple oxygen states. In particular

$$2CO_2 + S^{2-} + 2H_2O + h\nu \rightarrow 2CH_2O + SO_4^{2-} \tag{5.20}$$

And also

$$CO_2 + 4FeO + H_2O + h\nu \rightarrow CH_2O + Fe_2O_3 \tag{5.21}$$

These processes represent some kind of primitive photosynthesis considering the production of CH_2O.

5.4 Salinity

Paleontology and biochemistry together may yield fairly definite information, eventually, about the paleochemistry of seawater and atmosphere. Several less conclusive lines of evidence now available suggest that the composition of both seawater and

Table 5.1 Concentrations of the main dissolved species in the Archean and Modern ocean (in mmole/L) and their ratio.

Cl^-	B_r^-	I^-	SO_4^{2-}	Na^+	K^+	M_g^{2+}	Ca^{2+}	Sr^{2+}
Archean								
920	2.25	0.037	2.3	789	18.9	50.9	232	4.52
Present								
556	0.86	0.0005	28.7	477	10.1	54.2	10.5	0.09
Ratio								
1.65	1.16	74	0.28	1.65	1.87	0.94	22.1	50.22

atmosphere may have varied somewhat during the past, but the geologic record indicates that these variations have probably been within relatively narrow limits. A primary problem is how conditions could have remained so nearly constant for so long. This sentence is taken from the paper by William Rubey *Geologic History of Sea Water* published in 1951. It is rather surprising that in several decades the situation has not changed that much. Today we know that the sources of salt for the seawater are the weathering of continental crust and hydrothermal that is pouring from Mid-Ocean Ridges (MOR). It has been established that up to the Proterozoic (starting 2.5 Ga) the hydrothermal source was dominant and this conclusion has been reached through the examination of isotopic studies on Sr (strontium) and Nd (neodymium). ^{87}Sr is produced by the decay of rubidium ^{87}Rb which enrich the crust during solidification. The ratio $^{87}Sr/^{86}Sr$ increases in the crust compared to the values observed in the mantle. By comparing this ratio in the present sediments and those of the Archean values from about 0.715 to 0.703 are observed indicating an increasing contribution from the MOR source. It would be helpful for the evolution of chemical composition to study seawater inclusions in old rocks. These reveals the presence of NaCl (halite) together with calcium carbonate ($CaCL_2$). Table 5.1 reports a possible evolution in the composition of seawater.

This table shows that cation's concentration (Na^+, K^+, M_g^{2+}, etc.) has remained substantially the same except for calcium for the reasons we have seen before. In modern seawater, Br and I are depleted because they are fixed in plankton (they are metabolites for plankton).

5.4.1 *Salinity and the Age of the Ocean*

The problem of the constancy of salinity in the geologic time is strictly related to what we may call N^+ or salt cycle as shown in Fig. 5.5. We see that rivers are only one of the input to the ocean. The river source is made up by three different processes, the sea spray, the silicate weathering, and the chloride solution. The sea spray components are the aerosols produced by the evaporation of the droplets that

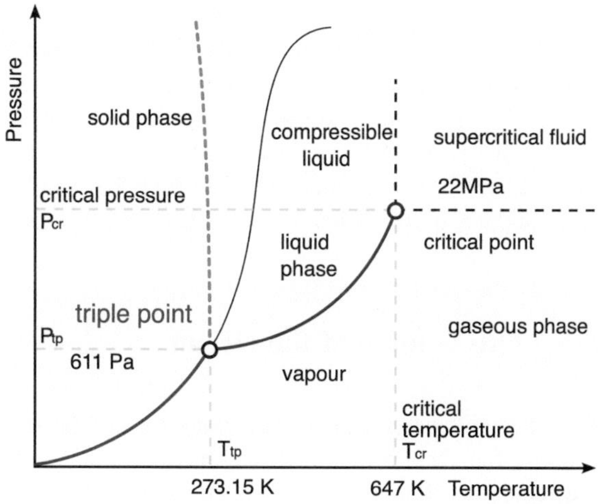

Fig. 5.5 The qualitative representation of the triple point and the critical point of water

are rained out. The silicate weathering and the chloride solution refer, respectively, to the erosion of silicate minerals and chloride minerals. The river source amounts to about 192 $Mt(megatons = 10^9\text{kg})year^{-1}$. Another major source of sodium ion is the sea floor sediments. The explanation of these sediments as a source is that most Na^+ are released from clay during prolonged burial as K^+ or Mg^{+2} is absorbed in its place. K^+ would be absorbed during prolonged burial of clay because it has larger ionic radius than Na^+. Similarly, Mg^{+2} would be absorbed by clay during prolonged burial because of its divalent charge, twice that of Na^+. This source amounts to 115 $Mt\ year^{-1}$. The third most important source is Na^+ water that seeps seaward through the continent. This source is quite uncertain but could amount to about 96 $Mt\ year^{-1}$. Other major sources include glacial silicates and hydrothermal vents. We can write a budget equation where the input rate in indicated as A_t while the output rate for each process must be proportional to the amount $M(t)$ so that the loss process may be written as $L_i = \beta M(t)$. The change in mass is then

$$\frac{dM}{dt} = A_t - \beta M(t) \tag{5.22}$$

which has the solution

$$M(t) = \frac{A_t}{\beta} - \left(\frac{A_t}{\beta} - M_0\right)\exp(-\beta t) \tag{5.23}$$

where M_0 is the initial mass. Assuming the initial mass to be $M_0 = 0$ we can find the time T necessary to build up the present amount of salt M_p

$$T = \frac{1}{\beta}\ln\left(\frac{A_t - \beta M_0}{A_t - \beta M_p}\right) = \frac{\tau}{x}\ln\left[\frac{1-(M_0 x/M_p)}{1-x}\right] \tag{5.24}$$

where τ is the filling time $\tau = M_p/A_t$ and $x = \beta M_p/A_t$. At this point knowing the total mass of Na^+ which is given by its density (10.76 g/kg) multiplied by the mass of the ocean (1.37×10^{21} kg) obtain $M_p = 1.47 \times 10^{19}$kg. The total $A_p \approx 4.57 \times 10^{11}$ kg/*year* and the filling time is then $M_p/A_p \approx 32.2 Myr$. Actually the correct age would be given by (5.24) where x is the ratio between the output and input. Even for lower values of x the time T remains of the order of tens of million years and it is quite evident that we need to know more about the history of the ocean.

5.5 The Volume of the Ocean

Wee have discussed how oceans were formed on Earth and have mentioned in the process that surface of continents may have changed. Now we ask if the volume of the ocean has changed with time. We will base our considerations on two papers by James Kasting but before we do that we need to refrain what is the Rayleigh number for porous media and what is the critical point for water.

The Rayleigh Number. We have defined the Rayleigh number as

$$Ra = \frac{g\alpha D^3 \Delta T}{\kappa \nu} \tag{5.25}$$

where g is the acceleration of gravity, α is the volume thermal expansion coefficient, h is the distance between the convecting plates, ΔT is the temperature difference, κ is the thermal diffusivity $\kappa = K/\rho C_p$ with ρ density, K is the conductivity of the fluid, and ν is the kinematic viscosity. In case of porous media it becomes

$$Ra = \frac{\alpha_w g C_{pw} \rho_w \Delta T P h}{\nu_w K_m} \tag{5.26}$$

where the variables with "w" refer to water and P is the permeability of rock, K_m thermal conductivity of water-rock medium. Equations (5.25)–(5.26) are the same because the porosity is measured in Darcy which correspond to m^2.

Critical Point. Figure 5.6 shows the familiar phase diagram for pure water. In the pressure-temperature diagram it is possible to see the *triple point* where all the three phases liquid. Gas and solid coexist. Then at much higher temperature and pressure the line separating the liquid and gas phase just disappears. This happens at a temperature of 647 K and a pressure of 22 megapascals or about 220 bars. At the critical point the properties of water change completely and similar to the gas, for example, become compressible, and transport of heat is much more efficient. In 1992, Kasting and Holms published a paper which started from the observation that near the oceanic ridge the temperature and pressure were similar to those of the critical point. They also considered the *Nusselt Number, Nu* which is defined as the ratio of the total heat flux (q) to the heat flux that could be maintained in the absence of convection ($K_m \Delta T/h$)

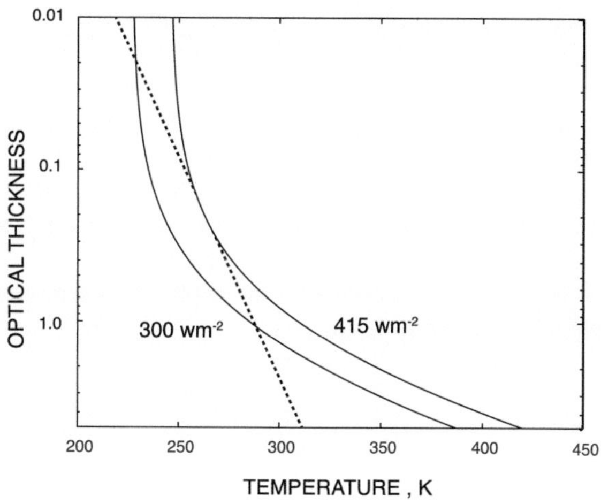

Fig. 5.6 Radiative equilibrium atmosphere. The dashed line is the optical thickness as a function of temperature. Also shown are the temperature profiles for 415 w/m^2 and 300 w/m^2

$$Nu = \frac{qh}{K_m \Delta T} \tag{5.27}$$

Several theories relate the Nusselt number to the Rayleigh number and in this case we adopt

$$\begin{aligned} Nu &\approx Ra/40 \quad Ra < 1000 \\ Nu &\approx 2.4 Ra^{1/3} \quad Ra > 1000 \end{aligned} \tag{5.28}$$

For a rock-dominated system K_m is essentially equal to the conductivity of the rock, so the Rayleigh number varies linearly with the ratio $R_w = -\alpha_w \rho_w C_{pw}/\nu_w$. This factor can be calculated showing that for pure water there is an extended region near and somewhat above the critical pressure where R_w, and hence Ra, can be more than 1000 times its low temperature value. (Indeed, R_w at the critical point itself is more than 10^6 times the low temperature value.) The enhanced value of R_w for a superconvective hydrothermal system implies that the depth to which the circulation can penetrate is substantially higher than for a cold, low-pressure system. If one assumes that the permeability and conductivity of the rock, the temperature difference ΔT, and the total heat flow q are invariant, and assume that $Nu \approx Ra^{1/3}$ then from (5.27)–(5.28) we have

$$\left(\frac{qh}{K_m \Delta T}\right)^3 = \frac{R_w g D \Delta T h}{K_m} \tag{5.29}$$

so that

$$h = g D \Delta T^4 K_m^2 \left(\frac{R_w^{1/2}}{q^{3/2}}\right) \propto \left(\frac{R_w^{1/2}}{q^{3/2}}\right) \tag{5.30}$$

This relation implies the constancy of $gD\Delta T^4 K_m^2$ and in this case the depth h would increase with R_w. We now can imagine a scheme which keeps constant the ocean volume once this has formed. In this case, we assume that the water is released from the mantle through the seafloor ridges. Initially outgassing would be larger than ingassing because the subducting crust would not be hydrated enough. As a matter of fact following (5.30) we know that in early Earth both R_w would be smaller than today while q could have been larger. Outgassing would remain dominant until enough water had filled the seafloor and submerged the ridge. At that point pressure would be near the critical point and the depth of hydration would be such that ingassing would balance outgassing.

Appendix

Cooling of the Earth

We refer to the paper by Sleep and Zahnle (2001) and trying to explain the numbers appearing there. They start by assuming an initial temperature for the Earth of 2300 K which would imply a surface flux $Q = \sigma T^4 \approx 1.6 \times 10^6$ wm^{-2}. If we assume a specific heat for the Earth of 10^6J kg^{-1} which is equivalent to a heath capacity of $6 \times 10^{24} \times 10^6 \approx 6 \times 10^{30}$ J. If the temperature would be 1500 K this would correspond to surface flux of 0.29×10^6 wm^{-2} that would dissipate all the heath in

$$\frac{6 \times 10^{30}}{2.9 \times 10^5 \times 3.15 \times 10^7 \times 5.14 \times 10^{14}} \approx 130\ year$$

On the other hand, if the dissipated flux is 150 wm^{-2} then the cooling time would be

$$\frac{6 \times 10^{30}}{150 \times 3.15 \times 10^7 \times 5.14 \times 10^{14}} \approx 2.5 \times 10^6\ year$$

The value of 150 wm^{-2} was calculated in a paper by James Kasting as the difference between the critical greenhouse threshold for a water-rich atmosphere and the heat supplied by the sun. The heat supplied by the Sun at that epoch was somewhat less ($\approx$70%) than the today flux (see the faint sun paradox). The greenhouse threshold in this case refers to the so-called *Komabayashi–Ingersoll limit* which is the minimum value of solar flux necessary to start a runaway greenhouse. We have examined that before but it is worth recalling that in a simple gray atmosphere the optical thickness is related to the partial pressure of vapor e

$$\tau = Ku = K\frac{ug}{g} = K\frac{e}{g} \tag{5.31}$$

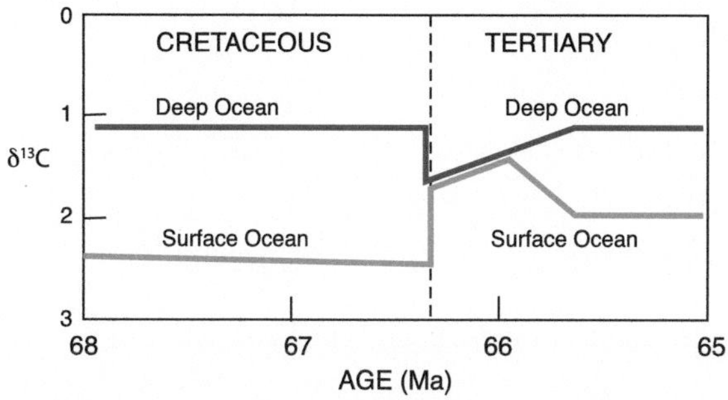

Fig. 5.7 Qualitative behavior of $\delta^{13}C$ between cretaceous and tertiary

where K the absorption coefficient, u is the mass of water vapor over unit surface so that ug it is just the partial pressure of water vapor. In case of saturation $e = e_s = 611\ exp[\beta(T - 273.15)]$ where $\beta = 0.067$. In this case, the optical thickness is just a function of temperature. On the other hand also the radiative temperature is function only of optical thickness and the solar flux F_0

$$\sigma T^4 = \frac{1}{2} F_0 \left(1 + \frac{3}{2}\tau\right) \tag{5.32}$$

We can change F_0 until the curve for $T = T(\tau)$ is tangent to the curve $\tau = \tau(T)$ as shown in Fig. 5.7. Values of the flux larger than about 400 wm^{-2} cause a runaway greenhouse. Actually it can be shown using not just a radiative model but a radiative-convective model that at the least for water vapor this large outgoing fluxes are not realistic and for the Earth the largest values is about 300 wm^{-2} as shown in the same Fig. 5.7. The reason is that above the tropopause for larger fluxes the conditions for saturation are met. In the figure the different regimes are clearly shown. Notice that the upper limit 300 wm^{-2} is about 70 wm^{-2} larger than the reduced incoming flux expected for that epoch.

The pH of the Ocean

Carbon dioxide dissolves in the ocean forming the weak carbonic acid ($CO_2 \cdot H_2O$) which dissociates in bicarbonate (HCO_3^-) and carbonate (CO_3^{2-})

$$\begin{aligned} CO_2(g) &\overset{H_2O}{\Longleftrightarrow} CO_2 \cdot H_2O \\ CO_2 \cdot H_2O &\Leftrightarrow HCO^- + H^+ \\ HCO_3^- &\Leftrightarrow CO_3^{2-} + H^+ \end{aligned} \tag{5.33}$$

with the equilibrium constants

$$K_H = \frac{[CO_2 \cdot H_2O]}{P_{CO_2}} = 3 \cdot 10^{-2}$$
$$K_1 = \frac{[HCO_3^-][H^+]}{[CO_2 \cdot H_2O]} = 9 \cdot 10^{-7} \tag{5.34}$$
$$K_2 = \frac{[CO_3^{2-}][H^+]}{[HCO_3^-]} = 7 \cdot 10^{-10}$$

The product of these equilibrium constant gives

$$K_1 \cdot K_2 \cdot K_H = \frac{[CO_3^{2-}][H^+]^2}{P_{CO_2}} \tag{5.35}$$

From the definition of pH

$$pH = -log[H^+] = log\left(\frac{K_1 \cdot K_2 \cdot K_H P_{CO_2}}{[CO_3^{2-}]}\right)^{1/2} \tag{5.36}$$

To evaluate $[CO_3^{2-}]$ we assume the water to be saturated with CaO_3 so we have

$$CaCO_3 \Leftrightarrow Ca^{2+} + CO_3^{2-}, \quad K_S = [Ca^{2+}][CO_3^{2-}] = 9 \times 10^{-7} M^2 \tag{5.37}$$

where Ca^{2+}=0.01 M. Substituting the numerical values we have pH = 8.53.

Urey (or Emiliani) Thermometer

Harold Urey published a paper in Science in 1948, *Oxygen Isotopes in Nature and in the Laboratory* where he found that ratios of the isotopes of an elements depend on the temperature. We consider two isotopes of oxygen ^{18}O and ^{16}O and define the variation

$$\delta^{18}O = \left[\frac{\left(\frac{^{18}O}{^{16}O}\right)_{sample}}{\left(\frac{^{18}O}{^{16}O}\right)_{standard}} - 1\right] \times 1000 \tag{5.38}$$

where "sample" refers to the sample examined and "standard" refers to some known water composition like the Vienna Standard Mean Ocean Water (VSMOW). This discovery had immediate application to find past temperature of the sea, as a matter of fact the calcium carbonate deposits reflect the oxygen composition of the sea when they are formed. Later on Nick Shackleton found that there was larger effect due to the accumulation of ice in the polar caps that we may call *glacial effect*. The ocean water evaporates and those vapors are enriched in lighter isotope (^{16}O) while the ocean water are depleted of the same isotope. The clouds transport water vapor over the polar regions where through snow precipitation they form the polar caps that are enriched in the lighter isotope. This effect was discovered by examining the Benthic foraminifera. These are marine amoebas that make fossilizable shells of calcite or cemented sediment grains. Because they reside at the ocean bottom they do not feel the temperature effect while are subject to the glacial effect. As an example, the first relation which connects the temperature to the isotopic composition is

$$T = 16.5 - 4.3(\delta_{carb} - \delta_{sea}) + 0.14(\delta_{carb} - \delta_{sea})^2 \tag{5.39}$$

The glacial effect could be used to obtain the ice volume. Maintaining constant the temperature in (5.33) (for example, 1°) it is possible to solve the equation for $\delta^{18}O_{sea}$ for the extreme values of $\delta^{18}O_{carb}$ (between maximum glacial and maximum interglacial) to estimate the amount of $\delta^{16}O$ stored in the ice. It is possible to calculate how much "light" water it is stored in the caps and consequently the variation in the sea level. It is found that $1^o/_{oo}\delta^{18}O$ corresponds to 100 m of ocean or $\Delta T \approx 5°$.

Strangelove Ocean

We want to mention another application of the isotope method which refers to the carbon isotopes. In 1985, Kenneth Hsu and Judith McKenzie published a paper about a carbon isotope anomaly in the planktic skeletons. They interpreted this anomaly as a sudden decrease in the biological activity of the ocean that coincides with the impact of the asteroid that wiped out the dinosaurs about 66 million years ago (Cretaceous/Tertiary boundary). At some point they affirm:... *The changes in ocean chemistry caused by temporary suppression of photosynthesis and catastrophic environmental pollution could have resulted in an ocean almost devoid of plankton, or a "Strangelove" ocean.* The name comes from the novel by Lloyd George translated in a movie by Stanley Kubrick. Let us start by defining the so-called $\delta^{13}C$.

$$\delta^{13}C = \left[\frac{\left(\frac{^{13}C}{^{12}C}\right)_{sample}}{\left(\frac{^{13}C}{^{12}C}\right)_{standard}} - 1 \right] \times 1000 \tag{5.40}$$

In the photosynthetic process organic matter tends to use the lowest isotope ^{12}C so that when there is a decrease in the biological activity there is a decrease on this ratio. A decrease was observed as schematically shown in Fig. 5.7 in the deep ocean. This would be in accord with the hypothesis of a complete cessation of primary production in the surface ocean. This would lead to a disruption of the isotopic gradient between the surface and the deep ocean. Organic carbon would be no longer exported from the surface water to deep waters. In 1991, Lee Kump proposed a simple model to explain the change in $\delta^{13}C$. It consists in two boxes (see Fig. 5.8) one representing the surface water that contains a mass M_s. This box receives carbon essentially from weathering

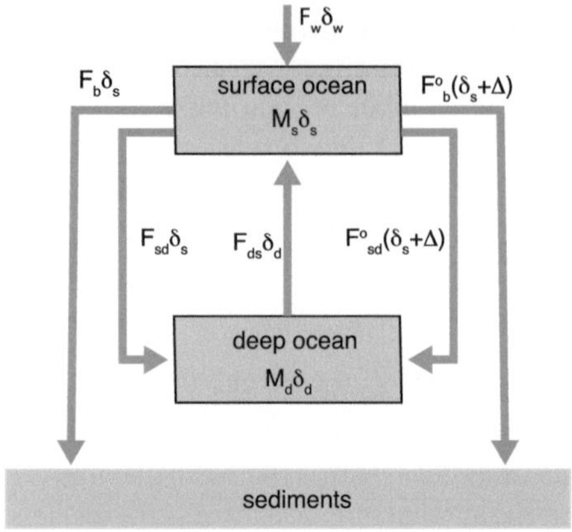

Fig. 5.8 Box model for the Strangelove ocean

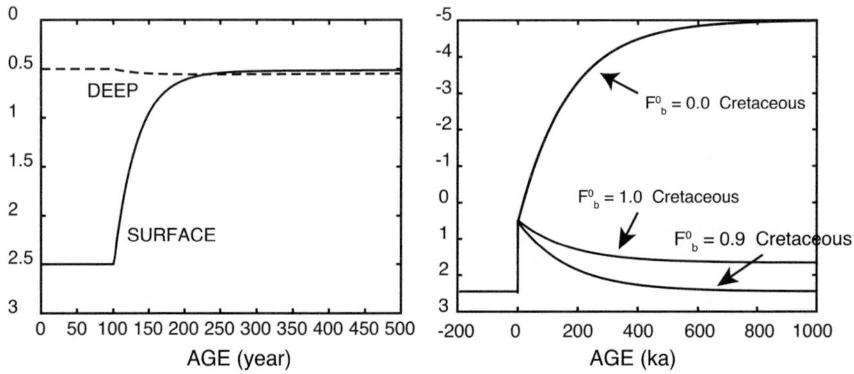

Fig. 5.9 On the left it is shown the short-term scenario while on the right the long-term scenario for three values of burial flux

and exchange matter with the deep ocean. In the figure F_{sd} is the flux and inorganic carbon which includes both the inorganic carbon transferred to the deep ocean and the carbonate sediments. F_{sd}^0 is the organic carbon flux and includes carbon which is oxidized to inorganic in the deep ocean. These fluxes are augmented to give the total organic F_b^0 and carbonate F_b carbon burial fluxes. The characterization of such fluxes in terms of their isotopic composition is accomplished by the product of the fluxes in mole of carbon per year and isotopic composition. By referring to the figure we can write some balance equation

$$
\begin{aligned}
\frac{dM_s}{dt} &= F_w + F_{ds} - (F_b + Fsd + F_b^0 + F_{ds}^0) \\
\frac{dM_d}{dt} &= F_{sd} + F_{sd}^0 - F_{ds}
\end{aligned}
\tag{5.41}
$$

And similar equations for the isotope mass balance

$$
\begin{aligned}
\frac{dM_s\delta s}{dt} &= F_w\delta w + F_{ds}\delta s - [(F_b + F_{sd})\delta s + (F_b^0 + F_{ds}^0)(\delta s + \Delta)] \\
\frac{dM_d\delta_d}{dt} &= F_{sd}\delta s + F_{sd}^0(\delta s + \Delta) - F_{ds}\delta d
\end{aligned}
\tag{5.42}
$$

where δs and δd are the isotopic composition of the surface ocean and deep ocean while Δ is the carbon assimilated by plants. The last two equations can be recast in those of isotopic composition

$$
\begin{aligned}
M_s\frac{d\delta_s}{dt} &= F_w(\delta w - \delta_s) + F_{ds}(\delta_d - \delta s) - (F_b^0 + F_{sd})\Delta \\
M_d\frac{d\delta_d}{dt} &= F_{sd}(\delta_s - \delta_d) + F_{sd}^0(\delta_s + \Delta) - \delta_d)
\end{aligned}
\tag{5.43}
$$

It is interesting to discuss the steady-state solutions which are obtained setting the time derivative to zero. They read

$$
\begin{aligned}
F_w &= F_b + F_b^o \\
F_d &= F_{sd} + F_{sd}^o \\
F_w\delta_w &= F_b\delta_s + F_b^o(\delta_s + \Delta) \\
F_{sd}\delta_s &= F_{sd}^o(\delta_s + \Delta) = F_{ds}\delta_d
\end{aligned}
\tag{5.44}
$$

Using the first and third equations, we have

$$
\delta_s = \delta_w - \Delta\frac{F_b^o}{(F_b^0 + F_b)}
\tag{5.45}
$$

While from the third we can find the gradient between the surface and the deep ocean

$$
\delta_s - \delta_d = -\Delta\frac{F_{sd}^o}{(F_{sd}^o + F_{sd})}
\tag{5.46}
$$

These equations show that the surface isotopic composition tends to the weathering composition in absence of the global organic burial rate ($F_b = 0$) and that the gradient disappears in the absence of the flux of carbon to the deep ocean ($F_{sd^0} = 0$). In the Lee Kump model, the Strangelove ocean is simulated by putting to zero the burial flux of organic carbon the flux of organic carbon F_b^0 from surface to deep ocean F_{sd}^0. It is quite instructive to give the values of the variables appearing in Eq. (5.36) as it is done in Table 5.2. It quite obvious from this table that there is a short-term response determined by the factor

$$
\frac{M_s}{(F_w + Fds)} \approx 30 years
$$

Table 5.2 The data for Eq. (5.36)

Fluxes 10^{13} mole/yr	Reservoir (10^{13} mole)	Isotopic value
$F_w = 0.02$	$M_s = 100$	$\delta_s = +2.5$
$F_b = 0.014$	$M_d = 3000$	$\delta_s = +0.5$
$F_b^o = 0.006$		$\delta_w = -5$
F_{sd}=2.76		$\Delta = -5$
$F_{sd}^o = 0.24$		
$F_{ds} = 3.0$		

Note F_w = weathering flux, F_b = carbonate burial flux, F_b^o = organic carbon burial flux, F_{sb} = surface to deep carbon flux, F_{sb}^o = organic carbon surface to deep flux, F_{ds} = deep to surface carbon flux, M_s = mass of surface reservoir, M_d = mass of deep reservoir δ_s, δ_d, δ_w are the isotopic composition of surface, deep, and river, while Δ is the photosynthetic fractionation factor

And a longer response determined by the term

$$\frac{M_d}{(F_{sd} + F_{ds}^o)} \approx 1000 years$$

if we run the model for a few hundred years with $F_b = Fsd^o = 0$ we observe the short-term response with the surface isotope which reaches the value of deep ocean. On the other hand for longer time scales the surface reaches the steady-state value δ_w. Notice that the only difference between the two programs is the conversion of the fluxes in C/kyr.

```
[t,x]=ode45('stranamore',[0 1000],[2.5 0.5]);
plot(t,x(:,1),t,x(:,2)), axis([-200 1000 -5 3])
set(gca, 'YDir','reverse')
function xdot=stranamore(t,x);
Fw=0.02;
Fb=0.014;
Fbo=0.;
Fsd=2.760;
Fsdo=0.;
Fds=3;
Ms=100;
Md=3000;
delw=-5;
delta=-25;
xdot=zeros(2,1);
xdot(1)=(1/Ms)*(Fw*(delw-x(1))+Fds*(x(2)-x(1));
xdot(2)=(1/Md)*(Fsd*(x(1)-x(2)));
```

References

Albarede, F.: Chemical archeoceanography. Chem. Geol. **548**, 119625 (2020)

Austin, S.A., Humphreys, D.R.: The sea's missing salt: a dilemma for evolutionists. In: Walsh RE, Brooks CL (eds) Proceedings of the second international conference on creationism, pp 17–33 (1990)

Catling, D.C., Kasting, J.F.: Atmospheric Evolution on Inhabited and Lifeless Worlds. Cambridge University Press (2017)

Genda, H.: Origin of Earth's oceans: an assessment of the total amount, history and supply of water. Geochem. J. **50**, 27 (2016)

Goldblatt, C., Watson, A.J.: The runaway greenhouse: implications for future climate change, geoengineering and planetary atmospheres. Phil. Trans. R. Soc. A **370**, 4197 (2012)

Grotzinger, P., Kasting, J.F.: New constraints on Precambrian Ocean composition. J. Geol. **101**, 235 (1993)

Halevy, I., Bachan, A.: The geologic history of seawater pH. Science **355**, 1069 (2017)

Holland, The geologic history of seawater, Treatise on Geochemistry, 6

Holland, H.D.: The Chemical Evolution of the Atmosphere and Oceans. Princeton University Press (1984)

Kasting, J.F., et al.: Paleoclimates, ocean depth, and the oxygen isotopic composition of seawater. Earth Planet. Sci. Lett. **252**, 82 (2006)

Kasting, J.F., Holms, N.G.: What determines the volume of the oceans? Earth Planet. Sci. Lett. **109**, 507 (1992)

Lee, C.A.: A framework for understanding whole-earth carbon cycling. In: Orcutt, B., Daniel, I., Dasgupta, R. (eds.), Deep Carbon: Past to Present. Cambridge University Press (2019)

Lowenstein, T.K. et al.: The geologic history OD seawater. Treatise on Geochemistry, 2nd edn. vol. 8, p. 569 (2014)

Morbidelli, A., et al.: Source regions and timescales for the delivery of water to the Earth. Meteoritics Planet. Sci. **35**, 1309 (2000)

Morse, J.W., MacKenzie, F.T.: Hadean ocean carbonate geochemistry. Aquat. Geochem. **4**, 301 (1998)

Nakashima, J., Hayashi, Y.Y., Abe, Y.: A study of the "Runaway Greenhouse Effect" with one dimensional Radiative-Convective equilibrium model. J. Atmos. Sci. **43**, 2256 (1992)

Pinti, D.L.: The origin and evolution of the oceans. In: Gargaud, M. et al. (eds.), Lectures in Astrobiology. Springer (2005)

Rothschild, L.J., Lister, A.M. (eds.), Evolution on Planet Earth. Ac. Press (2012)

Sleep, N.H., Zahnle, K.: Carbon dioxide cycling and implications for climate on ancient Earth. J. Geophys. Res. **106**, 1373 (2001)

Sleep, N.H., Zahnle, K., Neuhoff, P.S.: Initiation of clement surface conditions on the earliest Earth. Proc. Natl. Acad. Sci. U.S.A. **98**, 3666 (2001)

Tajika, E., Matsui, T.: Evolution of terrestrial proto-CO_2 atmosphere coupled with thermal history of the earth. Earth and Planet. Sci. Lett. **113**, 251 (1992)

Urey, H.C.: Oxygen isotopes in nature and in the laboratory. Science **108**, 489 (1948)

Chapter 6
Biogeochemical Cycles: The Rise of Oxygen (and Nitrogen)

The time has arrived to deal with the most important topic of this book: the evolution of oxygen in the atmosphere of the Earth. There is a huge amount of material on this subject and we decided that the most reasonable reference is the works of Heinrich Holland, James Kasting, and David Catling besides the classical approach of James Walker. It is quite clear that oxygen rise is related to the appearance of life on Earth but contrary to the popular belief the maintenance on the oxygen level is related to geochemistry and to the tectonic movements in the planet. The source of oxygen is clearly related to an imbalance between photosynthesis (which produces oxygen and organic matter) and respiration (which consumes oxygen) because part of the organic matter is subtracted to oxidation by burial. So it is clear that oxygen is related to the existence of life and to the cycling of carbon. As a matter of fact oxygen appeared on the Earth as soon as life originated produced by the *cyanobacteria* that contain chlorophyll and are capable of oxygen-producing photosynthesis.

6.1 The Regulation of Oxygen

It is worth to examine how the modern cycle of oxygen works. In Fig. 6.1, this cycle is shown in the essential terms. First of all, we can evaluate the amount of O_2 in the atmosphere which has a mass of $\approx 5.1 \times 10^{18}$ kg. The mass mixing ratio of oxygen is

$$\frac{\rho_{O_2}}{\rho_{atm}} = \frac{p_{O_2}}{p_{atm}} \frac{M_{O_2}}{M_{atm}} = 0.21 \frac{32}{29} = 0.23 \tag{6.1}$$

So the mass of molecular oxygen is $\approx 1.2 \times 10^6 Pg$ (1 Petagram = Pg = $10^{15} g$). In the same way, we can evaluate the mass of carbon dioxide in the atmosphere which has a mass mixing ratio of $\approx 4 \times 10^{-4}$ and the mass of $3000 Pg$ divided in 820 PgC and 2180 PgO.

G. Visconti, *Origin and Evolution of the Atmosphere and Ocean*,
https://doi.org/10.1007/978-3-032-02653-8_6

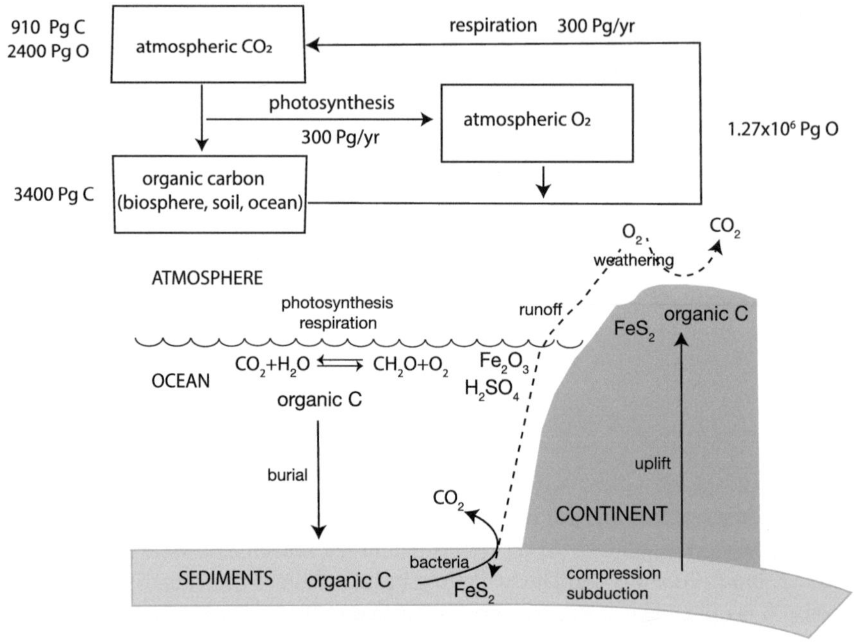

Fig. 6.1 The modern oxygen biogeochemical cycle

As shown in the figure the contribution of photosynthesis that can be represented by

$$CO_2 + H_2O \rightarrow CH_2O * O_2 \tag{6.2}$$

amounts to 0.273 Pg/year while the respiration

$$CH_2O + O_2 \rightarrow CO_2 + H_2O \tag{6.3}$$

Does not balance the photosynthesis but is roughly 0.1% less. This imbalance is due to the fact that part of the organic carbon resulting from photosynthesis is buried. In theory, the same imbalance should constitute the oxygen source. There are several reasons to object to this simple scheme. The total carbon can be divided into 700 Pg (biosphere), 2000 Pg in soil, and 1000 in the oceans for a total amount of about 3700 Pg. If photosynthesis were to stop all this carbon would oxidize consuming $4 \times 10^{18} \times 32/12 \approx 10^4\ PgO$ which represent less than 1% of the oxygen in the atmosphere. On the other hand, if respiration would stop then the conversion of all CO_2 into oxygen would produce $8.2 \times 10^{17} \times 32/12 \approx 2.2 \times 10^3\ Pg\ O$ again negligible with respect to the oxygen present in the atmosphere. Besides from the figure we see that the total oxygen buried is about 0.57 $Pg/year$ so that the residence time for O_2 is simply $1.2 \times 10^6/0.57 \approx 10^6\ years$. These considerations show that the biosphere cannot control the oxygen of the atmosphere which is rather counterintuitive. The control must involve longer time scales. The organic carbon deposited on the ocean floor is either buried or be processed by colorless bacteria in presence of sulfate and ferric iron (hematite Fe_2O_3) to form pyrite (Fe_2S) sediments

$$8SO_4^{2-} + 2(Fe_2O_3)_s + 15CH_2O + 16H^+ \rightarrow 4(Fe_2S)_s + 15CO_2 + 23H_2O \tag{6.4}$$

which combined with

$$15CO_2 + 15H_2O \rightarrow 15CH_2O + 15O_2 \tag{6.5}$$

gives

$$2(Fe_2O_3)_s + 16H^+ + 8SO_4^{2-} \rightarrow 4(FeS_2)_s + 8H_2O + 15O_2 \tag{6.6}$$

Now the discussion of balances is a little bit complicated. According to (6.5)–(6.5), each mole of carbon buried as CH_2O produces one mole of O_2 while !5/8 moles of O_2 are left for each mole of S buried as Fe_2S. Daniel Jacob in nice book gives 0.12 Pg C/year for the weathering of organic carbon and 0.05 Pg S/year for pyrite that would give a total weathering rate of

$$\begin{aligned}
&0.12Pg\ a^{-1}\ C = 0.01Pmol\ C = 0.01Pmol\ O_2 = 0.32Pg\ O\ a^{-1}\\
&0.05Pg\ a^{-1}\ S = 0.0016\ Pmol\ S = (15/8) \times 0.0016\\
&= 0.0029\, n\ Pmol\ O_2 = 0.093\ Pg\ a^{-1}\ O\\
&0.32 + 0.093 = 0.41Pg\ a^{-1}\ O
\end{aligned}$$

On the other hand, Catling and Claire give 0.48 Pg O/year for carbon and 0.32 Pg O/year for sulfur. Considering the amount of oxygen present the weathering rate reported by Jacob would give a lifetime of $1.2 \times 10^{21}/0.41 \times 10^{15} \approx 3$ million years while only 1.5 million year for the other data. These times may be responsible for short-term fluctuations while the reservoir of organic carbon in sedimentary rocks is enough to deplete the entire oxygen. The organic carbon and sulfur deposits are brought back to the surface where they are again oxidized in Fe_2O_3 and sulfuric acid closing the cycle (Fig. 6.1).

6.2 The History of Oxygen

Again you find an immense literature on the history of oxygen from the synthetic one by Lee Kump to the erudite one by Donald Canfield. The geologic evidence on the oxygen presence remains quite elusive but a few facts tell that the oxygen concentration rose from nothing to something very similar to the present mixing ratio ($\approx$21%) in a few million years. This would happen about 2.5 Ga ago during what is known *The Great Oxidation Event, GOE*. The evidence for this is the absence

Table 6.1 The possible composition of the atmosphere in late Archean (2.5 Ga) and middle Proterozoic (1 Ga). PAL = present atmospheric level of a gas

Species	Anoxic late Archean	Oxic middle proterozoic
Atmospheric O_2	<1 ppmv	0.2–2% by volume
Atmospheric CO_2	$\leq 4\%$	$\leq 4\%$
Atmospheric CH_4	$\approx 10^3$ ppmv	$\approx 10^2$ ppmv
Atmospheric N_2	<1.9 PAL	$\approx$10–20 PAL
Marine SO_4^{2-}	<0.2 mM	$\approx$1–4 mM

in older stratigraphic units of the so-called "red beds" , sedimentary rocks stained red by iron oxide. On the other hand, there was an abundance of lithified ancient soils that had lost their iron during weathering were found, reflecting the absence of oxygen in the weathering environment. The same uncertainty is found on the possible atmospheric composition before GOE. Table 6.1 taken from the Catling paper is a guess of this kind.

In the previous table, there are two terms "anoxic" and "oxic" that need some clarification. Oxygen is absent in an anoxic atmosphere whose composition is dominated by reducing gases like H_2 and CH_4 and their chemistry. Examples of reducing anoxic atmosphere are those of Jupiter, Saturn, Uranus, and Neptune with H_2 mixing ratio greater than 80%. Examples of oxic atmosphere are those of the Earth (21% O_2) or Mars (0.13% O_2). Archean atmosphere could have remained anoxic even if early oxygenic photosynthesis pumped O_2 into the atmosphere. Anoxia would have persisted if O_2 had been efficiently removed by a flux of reducing gases into the atmosphere. We can define an *oxygenation parameter* as

$$K_{oxy} = \frac{O_2 \text{ source flux}}{\text{non weathering } O_2 \text{ sink flux}} = \frac{F_{reductant_{burial}}}{F_{volcanic} + F_{metamorphic}} \tag{6.7}$$

In case this parameter is >1 we have an oxic atmosphere which becomes anoxic if the parameter is <1. In (6.7) volcanic stands for those which have volcanic origin while metamorphic are those processed by the crust. When $K_{oxy} < 1$, an atmosphere has very low O_2. When $K_{oxy} > 1$, an atmosphere is O_2 rich.

6.2.1 *The Sulfur Isotopes*

A very useful indication of the O_2 levels in the Archean comes from mass-independent fractionation (MIF) of sulfur isotopes, which suggests a level for oxygen less than 1 ppmv. There are four sulfur isotopes $^{32}S, ^{33}S, ^{34}S, ^{36}S$ and usually they are distributed according to a Mass-Dependent Distribution (MDF). Consequently, in MDF, ^{34}S is fractionated twice as much with respect to ^{32}S as is ^{33}S because the ^{34}S–^{32}S mass difference is twice that of ^{33}S–^{32}S. In terms of delta notation

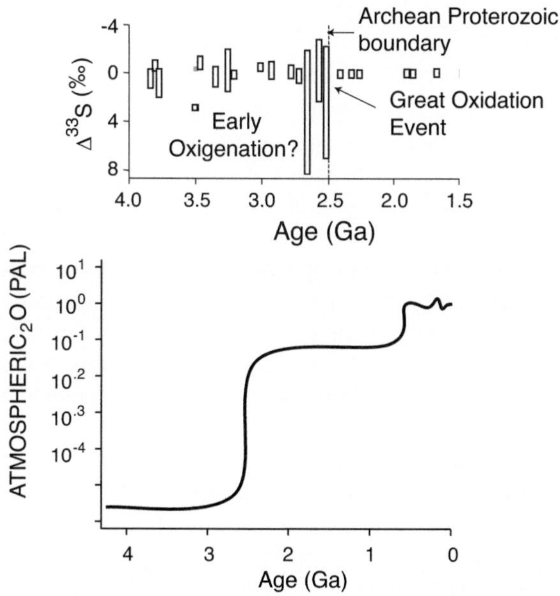

Fig. 6.2 In the upper part it is shown the behavior of $^{33}\Delta S$ while the lower part shows qualitatively the oxygen in terms of the present atmospheric level (PAL)

$$\delta^x S = \left[\frac{\left(\frac{^xS}{^{32}S}\right)_{sample}}{\left(\frac{^xS}{^{32}S}\right)_{standard}} - 1 \right] \times 1000 \tag{6.8}$$

variation in $\delta^{33}S$ usually about half those $\delta^{34}S$ in for MDF, which defines a straight-line relationship. In contrast, MIF is any fractionation that deviates from the linear relationship. That is

$$\begin{aligned} \delta^{33}S \approx \delta^{334}S \ \text{ is mass-dependent fractionation} \\ \delta^{33}S \neq \delta^{34}S \ \text{ is mass-independent fractionation} \end{aligned} \tag{6.9}$$

So for the MDF distribution we have $\delta^{36} \approx 2\delta^{34}$. A notation called "cap delta" is introduced to express the magnitude of the MIF and refers to a deviation from the MDF straight line as follows:

$$\begin{aligned} \Delta^{33}S = \delta^{33}S - 0.515\delta^{34}S \\ \Delta^{36}S = \delta^{34}S - 1.9\delta^{34}S \end{aligned} \tag{6.10}$$

Now if you plot this Delta as a function of time you get what it is shown in Fig. 6.2 where you can notice a discontinuity which is interpreted as a sign for the rise of oxygen. The reason for that is the fact that in the lab large MIF is produced in the dissociation of SO_2 and SO.

$$\begin{aligned} SO_2 + h\nu &\rightarrow SO + O \\ SO + h\nu &\rightarrow S + O \end{aligned} \tag{6.11}$$

The ultraviolet radiation in the early atmosphere that induces photolysis like (6.11) is in the wavelength range of 190–220 nm. Such ultraviolet photons can penetrate the troposphere in an anoxic atmosphere that lacks a stratospheric ozone layer. This was the first hypotheses to explain MIF. But then further work showed that other possibilities exist. In particular to preserve the MIF signature, three conditions are needed: very low atmospheric oxygen, sufficient sulfur gas in the atmosphere, and substantial concentrations of reducing gases. In particular, it was shown that in the latter condition the atmospheric methane level is the primary requirement for preserving MIF. Models show that the MIF signal was the direct result of a collapse in atmospheric methane levels. This loss of "greenhouse warming" is then invoked to explain the ensuing first major glaciation in Earth's history, perhaps of "snowball Earth." The MIF signal was best preserved in sediments when insoluble elemental sulfur S_8 was the product of the atmospheric photochemistry. The sulfur may be very important in explaining the rise of oxygen. The idea is based on the anaerobic oxidation of methane (AOM) which is performed by Archaea bacteria and can be expressed as

$$CH_4 + SO_4^{2-} \rightarrow HS^- + HCO_3^- + H_2O \tag{6.12}$$

Two factors determining the sulfate concentration in the early ocean, one is the increasing deposition from the growing continents and the other was the increasing decomposition of the organic deposit. The biogenic gases emitted would be O_2 and CH_4 that through reactions promoted by the microbiota could make methane according to the simplified scheme

$$\begin{aligned} 2H_2O + CO_2 &\xrightarrow{\text{photosynthesis}} 2CH_2O + 2O_2 \\ 2CH_2O &\xrightarrow{\text{fermentation+methanognesis}} CH_4 + CO_2 \\ &\text{Net } CO_2 + H_2O \rightarrow CH_4 + 2O_2 \end{aligned} \tag{6.13}$$

The result is that the fluxes of methane, Φ_{CH_4}, and oxygen, Φ_{O_2}, are in the ratio 2:1. Now we show with a simple model that the abundances of methane and oxygen are anti-correlated.

6.3 The Relation Between Methane and Oxygen

We follow the box model by Mark Claire et al. shown in Fig. 6.3. The fluid Earth is reduced to one box which represents the Atmosphere, Ocean, and Biosphere (AOB), while the solid Earth is represented by three boxes, the upper crust, the lower crust, and the mantle. F_B represents the flux of oxygen due to organic carbon burial, and

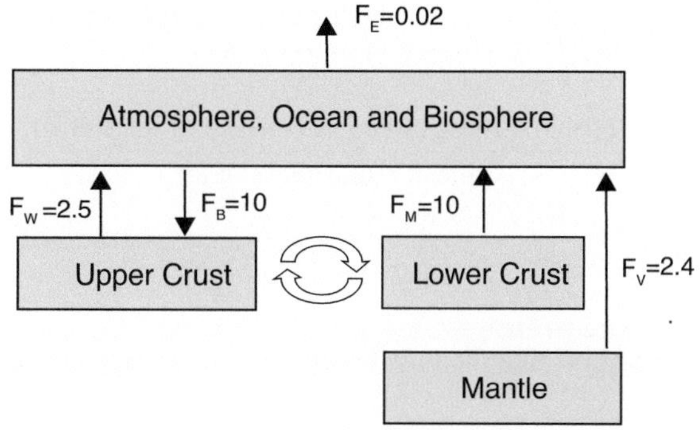

Fig. 6.3 The box model of Claire et al. (2006) for the calculation of the methane. The flux is expressed in equivalent Tmol $O_2\ yr^{-1}$

F_E, the flux of oxygen to the Earth as a whole due to hydrogen escape. The oxygen sinks F_V and F_M represent reducing gases (i.e., H_2, H_2S, $CO.CH_4$) from volcanic/hydrothermal and metamorphic/geothermal processes, respectively, while F_W is the oxygen sink due to oxidative weathering of continental rocks. Accordingly we can write the balance for O_2 and methane

$$
\begin{aligned}
\frac{d}{dt}[O_2] &= 2\Phi_{CH_4} + F_B - 2k_{eff}[O_2][CH_4] - k_{esc}[CH_4] - (F_V + F_M + F_w) \\
\frac{d}{dt}[CH_4] &= \Phi_{CH_4} - k_{esc}[CH_4] - k_{eff}[O_2][CH_4]
\end{aligned}
\tag{6.14}
$$

Here we need to be more detailed about fluxes. We start with the total hydrogen escape flux Φ expressed in $molecules$ cm^{-2} s^{-1}

$$
\Phi = k_c\ f_{total} = k_c(0.5 f_H + f_{H_2} + f_{H_2O} + f_{2CH_4} + f_{H_2S}...) \tag{6.15}
$$

where f is the mixing ratio for the different forms of hydrogen. In Eq. (6.15), we have described the hydrogen in terms of H_2 equivalents, although, depending on thermospheric conditions, hydrogen can also escape in atomic form. So k_c must be expressed as a flux itself and we take $k_c = 2.5 \times 10^{13}$ $H_2\, molecules$ cm^{-2} s^{-1}. At this point we can convert the $molecules$ cm^{-2} s^{-1} in Teramole per year using

$$
\begin{aligned}
\Phi &= \\
&\left[\frac{\text{flux} \times \text{surface of the Earth} \times \text{seconds in a year}}{\text{Avogadro number} \times 10^{12}\text{mol Tmol}^{-1}}\right] \times f_{total} \\
&= \left[\frac{2.5 \times 10^{13} \times 5.1 \times 10^{18} \times 3.15 \times 10^{7}}{6.02 \times 10^{23} \times 10^{12}\text{mol Tmol}^{-1}}\right] \times f_{total} \\
&= 6680\ f_{total}\ (\text{Tmol}\ \ H_2\ yr^{-1}) \\
&= 3340\ f_{total}\ (\text{Tmol}\ \ O_2\ yr^{-1})
\end{aligned}
\tag{6.16}
$$

Now we also need to evaluate how much CH_4 per bar we get in 1 bar N_2 atmosphere. If methane has a partial pressure ζ then

$$\frac{\text{(column mass of air)} \times \text{(surface of the Earth)}}{\text{mean molar mass of air}}$$
$$\frac{(10^5\ Pa/9.81) \times 5.1 \times 10^{14}}{\left(\frac{0.028+0.016\zeta}{1+\zeta}\right)(10^{12} mole Tmol^{-1})} = 1.8 \times 10^8 \text{ Tmol } CH_4\ bar^{-1} \tag{6.17}$$

We assume that the only contribution to escape is from methane so that

$$\begin{aligned} f_{total} = 2[CH_4] \Rightarrow F_E = k_{esc}[CH_4] \Rightarrow k_{esc} = 6680/18 \times 10^8 = \\ 3.7 \times 10^{-5} \text{ Tmol } O_2 \text{ equivalents } year^{-1} (\text{Tmol } CH_4^{-1}) \end{aligned} \tag{6.18}$$

Let us find the analytical solution to the (6.14) in the steady state $d[CH_4]/dt = 0$, $d[O_2/dt = 0$ so that

$$k_{esc}[CH_4] + F_B = F_E + F_B = F_v + F_M + F_W \tag{6.19}$$

Now according to (6.16) with a total hydrogen mixing ratio of 6 ppmv we have according to (6.16) an escape flux

$$F_{E,now} = 3340 \times 6 \times 10^{-6} \approx 0.02\ Tmol\ O_2\ yr^{-1}$$

Evaluating the different terms we have $F_{B,now} \approx 10 \pm 3.3\ Tmol\ year^{-1}$, $F_{V,now} \approx 2.4 \pm 1.8\ Tmol\ year^{-1}$, $F_{M,now} \approx 2.5\ Tmol\ O_2\ year^{-1}$. Also we have $F_W = k_w[O_2]^\beta$ and from (6.19) putting $F_E = 0$ we have

$$[O_2]|_{modern} = \left(\frac{F_{B,now} - (F_{V,now} + F_{M,now})}{k_W}\right)^{1/\beta} \tag{6.20}$$

where $k_w = 0.0065 Tmol^{2/3}\ year^{-1}$ for $\beta = 0.4$. We now consider the second of Eq. (6.14) in case $F_E \to 0$ we get

$$[CH_4]|_{modern} = \frac{\Phi_{CH_4}}{k_{eff}[O_2]} \tag{6.21}$$

The evaluation of k_{eff} corresponds to the synthesis of the problem of oxidation of methane. This process in the modern atmosphere starts with the reaction

$$CH_4 + OH \to CH_3 + H_2O \tag{6.22}$$

where the [OH] radicals are produced by two processes

$$
\begin{aligned}
H_2O + h\nu &\rightarrow OH + H \\
H_2O + O(^1D) &\rightarrow OH + OH
\end{aligned}
\tag{6.23}
$$

Both reactions require UV radiation and the oxidation process is detailed in the appendix. Here we assume that the process could be parameterized in terms of $[CH_4]$ and O_2 content following, for example, Zahnle et al. for a flux of $\Phi_{CH_4} = 400 Tmol^{-1}\ year^{-1}$, an oxygen level of 10^{-3} bar and we have $k_{eff} \approx 1.3 \times 10^{-7}\ Tmol^{-1}\ year^{-1}$ and so $[CH_4] \approx 400/(1.8 \times 10^5 \times 2.2 \times 10^{-7}$ we have a $\approx 18.000 Tmol$ of methane. On the other hand, 10^{-3} bar of oxygen corresponds to !.8 $\times$ 10^5 Tmol. There is another way to evaluate the oxygen level in the Archean. This is based on the fact that the level was so low that the oxidative weathering flux F_W was negligible. Substituting the escape flux $F_E = k_{esc}[CH_4]$ in (6.19) we have

$$
[CH_4]|_{Archean} = \frac{F_V + F_M - F_B}{k_{esc}} \tag{6.24}
$$

And using the conversion factor in (6.17) we get

$$
[CH_4]|_{Archean} = 1.5 \times 10^{-4} \left(F_{V,now} \frac{F_V}{F_{V,now}} + F_{M,now} \frac{F_M}{F_{M,now}} - F_{B,now} \frac{F_B}{F_{B,now}} \right) \tag{6.25}
$$

where $1/k_{esc} = 1/6680 \approx 1.5 \times 10^{-4}$ Now we write

$$
[CH_4]|_{Archean} = 1.5 \times 10^{-3}\text{bar} \times \left(\frac{3}{20} \left(\frac{F_M}{F_{M.now}} + \frac{F_V}{F_{V.now}} \right) - 1 \right) \tag{6.26}
$$

where we have substituted the values $F_B = 10$, $F_{M.now} = 1.5$, $F_{V.now} = 1.5$. Now consider that F_V and F_M are 7 times higher than today value while F_M may be 20 times higher than today (6.26) gives $[CH_4] \approx 3000\ ppmv$ (today is 1.7 ppmv). In the same way, we can estimate the oxygen mixing ratio by subtracting the second equation from the first in (6.14) to obtain

$$
0 = \Phi_{CH_4} - k_{eff}[O_2][CH_4] + F_B - (F_V + F_M + F_w) \tag{6.27}
$$

To find oxygen in the Archean we neglect F_W and obtain

$$
[O_2]_{Archean} = \frac{\Phi_{CH_4} + F_B - (F_V + F_M)}{k_{eff}[CH_4]} \tag{6.28}
$$

Using the previous estimate for methane (0.003 bar) the oxygen in the Archean amounts to 4×10^{-9} bar. The conclusion for now is that (6.21), (6.28) the abundance of methane and oxygen is anti-correlated, in the sense that if methane decreases then the atmosphere contains more oxygen. A complete run of the model we have discussed shows a sudden increase in oxygen following a minimum in the methane content as shown in Fig. 6.4.

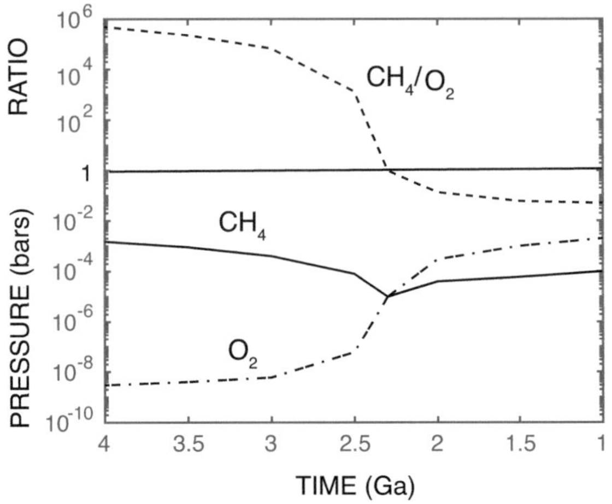

Fig. 6.4 The change in methane and oxygen pressures as a function of time from the present in Gigayear (Ga) as calculated from the model of Fig. 6.3

6.4 A Bistability Theory for the GOE

The sudden appearance of the GOE and the jump in the oxygen mixing ratio makes one think some bistability mechanism is acting in the atmospheric composition. As a matter of fact there is a paper by Colin Goldblatt and coworkers which theorize just that. The main point of this theory is oxidation of methane which is the inverse of (6.13)

$$CH_4 + 2O_2 \rightarrow CO_2 + 2H_2O \tag{6.29}$$

This reaction as shown in the appendix requires the presence of the hydroxil [OH] to start and that can be produced by UV radiation (see reactions (6.23)). However as the oxygen increases and ozone layer develops and becomes significant above about 10^{-3} PAL. Ozone partly shields UV radiation so that less radicals are produced and (6.29) slows down that is equivalent to reduce the oxygen sink. Goldblatt et al. made a review of the detailed models of methane oxidation and parameterized the oxidation rate of methane as a function of the ozone content with the relation

$$\text{oxidation rate} = \Psi M^{0.7} \tag{6.30}$$

where M is the amount of methane in moles and

$$\Psi = 10^{(a_1\psi^4 + a_2\psi^3 + a_3\psi^2 + a_4\psi + a_5)} \tag{6.31}$$

With $\psi = log O$ (O in moles), and $a_1 = 0.003$, $a_2 = -0.1655$, $a_3 = 3.2305$, $a_4 = -25.8343$, $a_5 = 71.5398$. The fit (6.31) is shown in Fig. 6.5 and it is interesting to note the maximum in Ψ around 10^{14} mole. At this point we need to fix the methane amount which results from this source r balanced by its loss sM so that $M = (r/s)$.

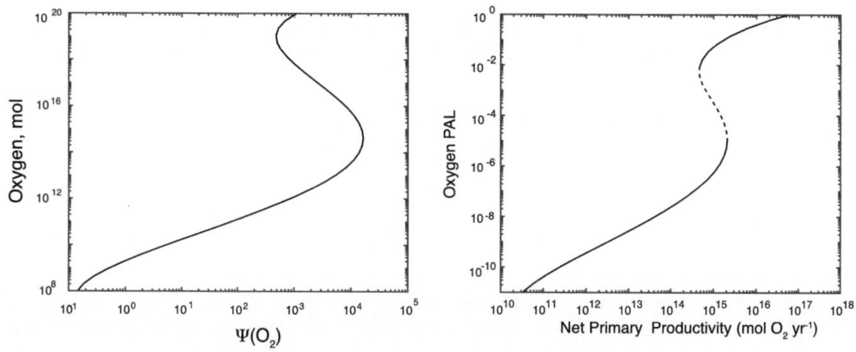

Fig. 6.5 On the left it is shown the oxidation parameter Ψ as a function of the oxygen moles while on the right it has shown the primary productivity. The dashed portion is the instability region

The reductant input r is modeled following the reaction in which ferrous iron acts as electron donor

$$4Fe^{2+} + CO_2 + 11H_2O + h\nu \rightarrow 4Fe(OH)_3 + CH_2O + 8H^+ \tag{6.32}$$

This is called anoxic photosynthesis and the resulting $Fe(OH)_3$ (ferrous hydroxide) is precipitated as Banded Iron Formation (BIF). The reductant input is another number to fix. In the above scheme for each mole of Fe there is 1/4 mole of oxygen. According to Holland the deposition scheme is a little bit different

$$6FeO + O_2 \rightarrow 2Fe_3O_4 \tag{6.33}$$

So that 6 moles of Fe require 1 mole of O_2. The flux of iron is evaluated between 5×10^{11} and 3×10^{12} $moles\ Fe\ yr^{-1}$ so r may change between 1.25×10^{11} and 7.5×10^{11} $moles\ O_2\ equiv\ yr^{-1}$. The loss of methane is attributed entirely to escape following the scheme

$$\begin{array}{c} CH_4 + h\nu \rightarrow 4H(\uparrow \text{space}) + C \\ C + O_2 \rightarrow CO_2 \\ \hline CH_4 + O_2 + h\nu \rightarrow 4H(\uparrow \text{space}) + CO_2 \end{array} \tag{6.34}$$

In this case, it is assumed that loss of methane is sM with $s = 2.03 \times 10^{-5}\ year^{-1}$ which corresponds to a lifetime of 50,000 years. To find the oxygen abundance we have to fix the source. We indicate with N the primary productivity from oxygenic photosynthesis. The oxygen flux from the biosphere is given by $\Omega_2 N$ where Ω_2 takes into account the oxygen lost by heterotrophic respiration (RH) and by methanotrophs. The RH represents the carbon losses from decomposition of litter and soil organic matter by microorganisms. In the same way, organism metanotrophs need photosynthetic oxygen to use CH_4. It is assumed that the necessary oxygen is taken from the primary productivity so that $\Omega_2 = (1-\delta)(1-\gamma)$ where the fraction of methane produced which is consumed by methanotrophs is $\delta = O/(d_\delta + O)$ with $d_\delta = 2.73 \times 10^{17}$ mol while the fraction consumed by heterotrophic respirers is $\gamma = O/(d_\gamma + O)$ with $d_\gamma = 1.36 \times 10^{19}$ mol. At this point, we can solve the equation

$$\Omega_2 N = \Psi \left(\frac{r}{s}\right)^{0.7} \tag{6.35}$$

In Fig. 6.5 N it is shown as a function of the oxygen level. We see that up to roughly 10^{-5} PAL the productivity grows with increasing PAL then up to about 10^{-2} PAL the productivity decreases with increasing oxygen and this region (shown in dashed line) is a region of instability. It can be understood because a decrease in productivity should correspond to a lower oxygen consumption while the solution shows an increase. The situation is that of two stable equilibria one with oxygen below 10^{-5} PAL and the other with high oxygen content ($>10^{-2}$ PAL) with the system jumping from one to another. The bistability theory has been reported because it is suggested by the apparent jumping of the GOE.

6.5 A Theoretical Framework?

Geophysics (it is what we are talking about) is fascinating and very useful field of study. People who practice in it sometimes suffer from an inferiority complex with respect to those physicists with costly and immense labs who practice what it is known as reductionism: to explain everything with an equation. In our case, some people also tried a similar approach for atmospheric evolution based mainly on the *redox budget*. The term "edox budget" is used to describe the balance between chemical reductants (electron donors) and chemical oxidants (electron acceptors). Earth's present atmosphere is highly oxidized because its second most abundant gas, O_2, is a powerful electron acceptor. Reduction and oxidation processes must always be balanced so that if one compound is oxidized the other must be reduced. For example, in the photosynthetic process

$$CO_2 + H_2O \rightarrow CH_2O + O_2$$

In this case, CO_2 is reduced (gain electrons) while oxygen gives up electrons to form O_2. When 1 mol of organic material (CH_2O) is buried 1 mol of O_2 is released. In the early Earth oxygen appears as a trace constituent so it makes more sense to keep track of reduced species like H_2 which is related to oxygen by the process

$$2H_2 + O_2 \rightarrow 2H_2O$$

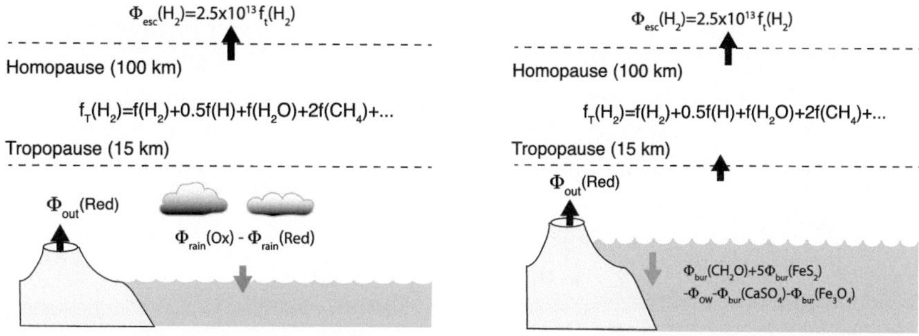

Fig. 6.6 The atmospheric redox budget (on the left) and the global redox budget (on the right)

where for each mol of oxygen 2 mol of hydrogen are consumed. For hydrogen we need to make an atmospheric budget as well as a global budget which includes also the ocean. Figure 6.6 shows what we intend for atmospheric redox budget: the reductant gases emitted from volcanoes are lost in the exosphere by escape.

According to general opinion we fix the present escape flux of hydrogen as $\Phi_{out}(H_2) = 2.4 \times 10^{12}$mol yr^{-1}. Notice that if we want to use another unit

$$1\text{mol } yr^{-1} = \frac{6 \times 10^{23}}{3.15 \times 10^7 \times 5.14 \times 10^{18}} = 0.00374 \text{ molecules } cm^{-2}s^{-1}$$

where the Avogadro number (6×10^{23}) has been divided by the seconds in a year and the Earth surface in cm^2. If we assume that hydrogen recombines with oxygen each mol of hydrogen that will escape will leave 0.5 mol of O_2 behind. This means that oxygen will be produced at a rate of 1.2×10^{12}mol yr^{-1}. It is interesting to note that the present amount of oxygen in the atmosphere ($\approx 3.1 \times 10^{19}$ mol would be produced in $3.1 \times 10^{19}/1.2 \times 10^{12} \approx 2.6 \times 10^7$ years. As we have seen a few chapters back the escape of hydrogen depends on the hydrogen mixing ratio at the homopause ($\approx$100 km) and is given by

$$\Phi_l(H) \approx 2.5 \times 10^{13} f_T(H) \text{ atoms cm}^{-2} \text{ s}^{-1} \tag{6.36}$$

where $f_T(H)$ is the mixing ratio for hydrogen at the homopause defined as

$$f_T(H) = f_H + 2f_{H_2} + 2f_{H_2O} + 4f_{CH_4} + ... \tag{6.37}$$

It is convenient to express the hydrogen flux in terms of H_2 and from (6.37) the major contribution is expected from water vapor whose mixing ratio at the stratosphere is about 3 ppm so that based on (6.36) the total flux is

$$\Phi(H_2) \approx (2.5 \times 10^{13}) \times (3 \times 10^{-6}) = 7.5 \times 10^7 \text{ molecules cm}^{-2}\text{s}^{-1} \tag{6.38}$$

The flux of oxygen is half this value that is 3.75×10^7 which converted in mol gives $\approx 1.0 \times 10^{10}$ mol O_2 per year that is 120 times less than the abiotically produced oxygen. In the early Earth there was enough oxygen that could react with the outgassed hydrogen that so was lost in space. We can obtain the mixing ratio of hydrogen if it was escaping at the diffusion limit. We use balance (6.36) with the outgassing rate Φ_{out}

$$\begin{aligned} \Phi_l(H) &\approx \Phi_{out} \\ 2.5 \times 10^{13} f(H_2) &\approx 1 \times 10^{10} \end{aligned} \tag{6.39}$$

And we obtain $f(H_2) \approx 400$ ppmv that is much larger than the present one (0.55 ppmv). A more rigorous discussion of the redox state should include also other production processes for hydrogen like

$$CO + H_2O \rightarrow CO_2 + H_2 \tag{6.40}$$

Or from outgassed hydrogen sulfide H_2S

$$H_2S + 2H_2O \rightarrow SO_2 + 3H_2 \tag{6.41}$$

In this case for each molecule of H_2S are produced 3 H_2 that could become 4 if SO_2 is further oxidized

$$SO_2 + 2H_2O \rightarrow H_2SO_4 + H_2 \tag{6.42}$$

In the last case, sulfuric acid can be regarded as a surrogate for gypsum which means that for the burial of one mol of gypsum one mol of H_2 is produced. There are also photochemical processes which produce H_2 like

$$2H_2O \rightarrow H_2O_2 + H_2 \tag{6.43}$$

where hydrogen peroxide (H_2O_2) can react with ferrous iron and produce again hydrogen

$$4FeO + H_2O_2 \rightarrow H_2 + 2Fe_2O_3 \tag{6.44}$$

However the major iron-bearing mineral is magnetite (Fe_3O_4) whose formation also releases hydrogen

$$3FeO + H_2O \rightarrow Fe_3O_4 + H_2 \tag{6.45}$$

Reactions (6.42)–(6.43) can also proceed from right to left and in this case H_2O_2 and H_2SO_4 act as oxidants with an equivalent H_2 which is negative -1. Finally burial of pyrite (FeS_2) is also a sink for hydrogen

$$4SO_2 + 10H_2 + 2'FeO' \rightarrow 2FeS_2 + 10H_2O \tag{6.46}$$

where 'FeO' stands for ferrous iron. However the details of such processes are illustrated in the appendix. The redox budget can be written as

$$\Phi_{out}(Red) + \Phi_{rain}(Ox) = \Phi_{esc}(H_2) + \Phi_{rain}(Red) \qquad (6.47)$$

The term Φ_{rain} refers to those processes where the compound is dissolved in water and is rained out like the case of formaldehyde (H_2CO))

$$CO_2 + 2H_2 \rightarrow H_2CO + H_2O \qquad (6.48)$$

In (6.45), we have

$$\Phi_{out}(Red) = \Phi_{out}(H_2) + \Phi_{out}(CO) + 4\Phi_{out}(CH_4) + 3\Phi_{out}(H_2S) \qquad (6.49)$$

where $\Phi_{out}(i)$ is the outgassing flux for species (i), $\Phi_{out}(Red)$ is the total outgassing flux of reduced species in units of H_2 equivalents, $\Phi_{rain}(Ox)$ is the rainout rate of photochemically produced oxidants, and $\Phi_{rain}(Red)$ is the rainout rate of photochemically produced reductants. In Eq. (6.47), those compounds which are neutral with respect to the oxidation state are not included . Kasting writes the redox with a more explicit equation

$$\begin{aligned}\Phi_{out}(red) + \Phi_{OW} + \Phi_{burial}(CaSO_4) + \Phi_{burial}(Fe_3O_4)\\ = \Phi_{esc}(H_2) + 2\Phi_{burial}(CH_2O) + 5\Phi_{burial}(FeS_2)\end{aligned} \qquad (6.50)$$

where the various terms read
$\Phi_{out}(red)$ = Total outgassing of reduced volcanic and metamorphic gases.
Φ_{OW} = Rate of oxidative weathering of the continents and seafloor.
$\Phi_{burial}(CaSO_4)$ = Burial flux of gypsum and anhydrite.
$\Phi_{burial}(Fe_3O_4)$ = Rate of anaerobic iron oxidation.
$\Phi_{esc}(H_2)$ = Rate of hydrogen escape to space.
$\Phi_{burial}(CH_2O)$ = Burial rate of organic carbon in sediments.
$\Phi_{burial}(FeS_2)$ = Burial rate of pyrite in sediments.
Now we can reformulate (6.7) by writing

$$K_{OXY} = \frac{F_B}{F_V + F_M} = \frac{2\Phi_{burial}(CH_2O) + 5\Phi_{burial}(FeS_2)}{\Phi_{out}(red) + \Phi_{burial}(Fe_3O_4)} \qquad (6.51)$$

The difference with respect to (6.7) is that we added the burial term of pyrite in the numerator and the burial of magnetite in the denominator. As before when $K_{OXY} > 1$ we have an oxidized atmosphere and when $K_{OXY} < 1$ we have a reduced atmosphere. It is worth to summarizes these fluxes as done by Kasting. After we know the values of the different terms we can resume the discussion why hydrogen escape did not cause the atmosphere to be oxygenated in few million years. We reach the conclusion that the present oxygen in the Earth's atmosphere would be build up in about 26 million years while geologic evidence shows that it took up to 700 million years to

Table 6.2 Modern global redox budget

H_2 source	Rate (10^{12} mol/yr)	References
$\Phi_{out}(red)$	4.8 ± 3.6	Holland (2002)
Φ_{OW} (continent)	30 ± 8	Holland (2002)
Φ_{OW} (seafloor)	2 ± 1	Sleep (2005)
$\Phi_{burial}(CaSO_4)$	1.0	Lasaga et al. (1985)
$\Phi_{burial}(Fe_3O_4)$	0.4 ± 0.2	Sleep (2005)
Total	38.1 ± 8.8	
H_2 sinks		
$\Phi_{esc}(H_2)$	0.04 ± 0.01	Walker (1977)
$2\Phi_{burial}(CH_2O)$	20 ± 6.6	Holland (2002)
$5\Phi_{burial}(FeS_2)$	10 ± 5	Holland (2002)
Total	30.1 ± 8.3	

establish an oxygenated atmosphere after the advent of photosynthesis. Kevin Zahnle and coworkers have some different ideas how to interpret such data. They start by considering the present amount of oxygen in the atmosphere ($\approx 0.31 \times 10^{20}$) and as from Table 6.2 consider that only burial of carbon would only 2 Myrs to build it up. The entire reservoir of oxygen in the continents ($\approx 5 \times 10^{20}$mol) would be build up in just 25 Ma. Other reservoirs like reduced carbon in continents ($\approx 1, 2 \times 10^{21} mol$) or oxidized continental basalt ($\approx 2 \times 10^{21} mol$) would employ relatively short time to fill out. The conclusion is that oxygen could raise in the atmosphere only after reservoirs were build up. The key point is to compare the flux of volcanic gas in the atmosphere with the rate of burial of carbon. If the fluxes of H_2 and CO exceed the carbon burial rate there is enough H_2 and CO to consume all the oxygen produced by the burial and in that case the atmosphere remains reduced. If the flux or reductants fall below the burial rate at this point the atmosphere starts to be oxygenated. The role of hydrogen escape is that to dissipate excess hydrogen which otherwise would go to make the continent more reduced. The hydrogen escape is fundamental to establish the direction of evolution. The authors conclude that the million of years that elapsed between the beginning of oxygen production by photosynthesis and the rise in the atmosphere were employed to oxidize the continents. Notice that there is a basic difference between the hydrogen escape that waste definitively H_2 and the role of carbon which is buried and resurfaced to consume an equivalent quantity of oxygen. Notice that in Table 6.2 the escape flux is decreased to 4×10^{10} mol/yr which is 250 times less than the previous value of 10^{13} mol/yr. This flux as shown in the appendix would evaporates 400 m of water in just 1Gyr and in the sale period builds 40 bar of oxygen.

6.5.1 *The Global Redox Budget*

We will refer mainly to the Kasting (2013) paper and utilize (6.51). In that equation we have neglected the burial of sulfate because it requires an excess of oxygen, as well as oxidative weathering and hydrogen escape. Then if we substitute the numbers in (6.51) we get

$$K_{OXY} = \frac{20+10}{4.8+0.4} = \frac{30}{5.2} = 5.8 \tag{6.52}$$

That means the present atmosphere is oxidative. We can now attempt some number for the Archean time. Here we get the usual dance with numbers and the uncertainties reign supreme. Most of the numbers refer to the Heinrich Holland model that we illustrate in the next paragraph but the immediate conclusion is that while the source rate would stay the same (4.8+0.4) together with the burial rate of carbon the burial of pyrite could have halved so that we should have

$$K_{OXY} = \frac{20+5}{4.8+0.4} = \frac{25}{5.2} = 4.8 \tag{6.53}$$

That just does not satisfy the expectation of a reducing atmosphere. So the final idea is that the burial of carbon is reduced by five times while the $\Phi_{out}(red)$ was probably doubled. This is consistent with a higher outgassing rate of reductant because of the hotter solid Earth. In this case

$$K_{OXY} = \frac{4+5}{9.6+0.4} = \frac{9}{10} = 0.9 \tag{6.54}$$

This leaves the time it finds and is a justification to write more papers in the subject.

6.6 The GOE and Lomagundi Event (LE)

Together with the Great Oxidation (GOE) another geologic event known as the Lomagundi Event (LE) sign the Archean Earth. During this event carbon isotope ratios of marine carbonates increased by up to ≈10‰, both events occurred around 2.3–2.0 billion years ago (Ga). As usual in presence of scarcity of data an avalanche of theories have been published. As an example here we report one advanced some times ago by James Eguchi and coworkers. According to this hypothesis around 2.5 Ga ago a major tectonic transition increased CO_2 emissions and consequently burial of carbonate and organic C via enhanced weathering and nutrient delivery to oceans. We should remember that photosynthesis started probably half billion year before the GOE while LE could have followed some 100 million year later. The preferred theory of GOE may be attributed to Holland that formulated it in 2009. The rise of oxygen is attributed to a decreased oxygen sink. However this does not explain why the C isotope fluctuation was observed. Contemporary to the enhanced emission in the same period Earth experienced the first glaciation (due probably to the decrease in methane) and the build up of the subaerial continents. Both events increased the deposition of hydrated sediments to trenches, lubricating the subduction interface and enhancing plate tectonic activity, which would have led to enhanced volcanic CO_2 emissions. An increase in tectonic activity is consistent with GOE that follows the photosynthesis activity as a matter of fact the production rate of oxygen depends on the flux of the organic carbon burial

$$F_{org} = f_{org} \times F_{tot} \tag{6.55}$$

where F_{org} is the flux of organic C burial and F_{tot} is the flux of total carbon burial (carbonate + organic C) while f_{org} is the fraction of carbon buried as organic. Relation (6.55) means that the flux of organic carbon may increase if the total flux increases with the fraction staying constant. At the beginning of photosynthesis the weathering flux was low due to the lack of subaerial continents. A change in tectonic activity could result in a change in F_{tot} and also explains the delay of GOE. On the other hand to explain the δC excursion we have to invoke a change in f_{org} because of the relationship

$$\delta^{13}C_{in} = \delta^{13}C_{carb}(1 - f_{org}) + \delta^{13}C_{org} f_{org} \tag{6.56}$$

where $\delta^{13}C_{in}$ is $\delta^{13}C$ of atmospheric inputs of CO_2, and $\delta^{13}C_{org}$ is $\delta^{13}C$ of organic C. The proposal of Eguchi is that the increase $\delta^{13}C_{carb}$ of the LE could have resulted from increased $\delta^{13}C_{in}$ rather than increased f_{org}. On the other hand, $\delta^{13}C_{in}$ depends on a number of sources

$$\delta^{13}C_{in} = \frac{F_{OIB}}{F_{tot}}\delta^{13}C_{OIB} + \frac{F_{MORB}}{F_{tot}}\delta^{13}C_{MORB} + \frac{F_{arc}}{F_{tot}}\delta^{13}C_{arc} \tag{6.57}$$

where F_i is the flux at the different volcanic settings with the relative $\delta^{13}C$. In particular, OIB stands for Ocean Island Basalt, MORB for Mid-Ocean Ridge Basalt and “arc” refer to volcanic arc. Increasing $\delta^{13}C_{in}$ by increasing $\frac{F_{arc}}{F_{tot}}$ is possible if $\delta^{13}C_{arc}$ is high, which is most easily explained by a heavy contribution of carbonates to arc CO_2 emissions. The model developed consists in six different reservoirs of carbon three which are located in the mantle. Both carbonates and organic carbon are deposited at the bottom of the sea and subducted in the Earth interior. Following subduction, organic C metamorphoses to graphite, which is more refractory than carbonate. At subarc locations carbonates are preferentially released with respect to graphitized organic C increasing $\delta^{13}C_{arc}$ to values greater than -5 ‰. Graphitized organic C may be transported deep into the mantle and released in oceans land volcanoes with lower values of $\delta^{13}C < -5$‰. This model may explain why GOE and LE are almost contemporaries. A change in tectonic stile produces a larger emission of carbon dioxide accompanied by a larger deposition of carbon and the oxygen production. The rapid emergence of subaerial continents15 or global glaciation proposed to have occurred in the same period may have facilitated such emission by a larger deposition of hydrated sediments.

6.7 And Now What About Nitrogen?

In most texts on the atmospheric evolution, the main topic is oxygen origin and the evolution, the reason being that we survive on Earth because of oxygen and the same element is associated with the origin of life. However, the same importance should be given to nitrogen which makes up almost 80% of the Earth atmosphere. Possibly there is a basic difference between the two elements, in one case (oxygen) it originated from photosynthesis that is with life while in the other (nitrogen) it was probably delivered directly from the material which made up the Earth, for example, carbonaceous chondrites. It is further observed that nitrogen is rather chemically inert and its utilization by living matter require some preliminary transformation. But the same is true for oxygen which is quite inert and must use other chemical (OH for example) to do its job. We will familiarize with those nitrogen processes which make it available and then compare the early biogeochemical cycle to the present one which we report in the appendix. Let us start with evaluating the amount of N present on Earth. It is rather surprising that the largest portion of nitrogen is found in the atmosphere, 3.9×10^9 Tg of N_2 corresponding to 5.6×10^{20} moles of N. The ocean contains about 8.0×10^5 Tg while the solid Earth may hold the same amount of the atmosphere as shown in Fig. 6.8 adapted from the work of Li. It is estimated that the crust contains half the amount of the atmosphere (1.8×10^9 Tg) mostly as NH_4 while the mantle is estimated to contain 2.5×10^9 Tg). It is interesting to notice the isotopic value $\delta^{15}N(‰) = [(^{15}N/^{14}N)_{sample}/(^{15}N/^{14}N)_{standard} - 1)] \times 1000$ which ranges between 0 ‰ in the atmosphere to +6 ‰ in the crust down to −40 ‰ for diamonds in the mantle. These ranges of values can be related to the formation process of the Earth considering that in the solar wind these values are of the order of −400 ‰ while reach +1000‰ in comets. In general, the isotopic value is related to biological activity because living organisms apparently prefer the lighter isotope. However, the inclusion processes are so complex that is hard to derive a correlation between the biological activity and the isotopic composition. One of these problems is related to the burial flux of nitrogen into the sediments. Following the origin of life ($\approx$3.8 Ga ago) about 3Ga ago organic material was produced by anoxygenic photosynthesis where bacteria store light energy by phototrophic process in ATP (Adensosine Triphosphate molecules), without the production of oxygen. The burial rate was roughly 20% of present rate. Oxygenic photosynthesis started just before GOE raising slightly the burial until the end of GOE. The productivity and burial rate increased to 80% with the appearance of the first eukaryotic algae about 1.5 Ga ago. The most compelling indication on the change in the partial pressure of N_2 in the atmosphere could be the requirement of the higher pressure to broaden the absorption line to enhance the greenhouse that would compensate the lower luminosity of the Sun. We will discuss this aspect in detail in the next chapter when we look at the

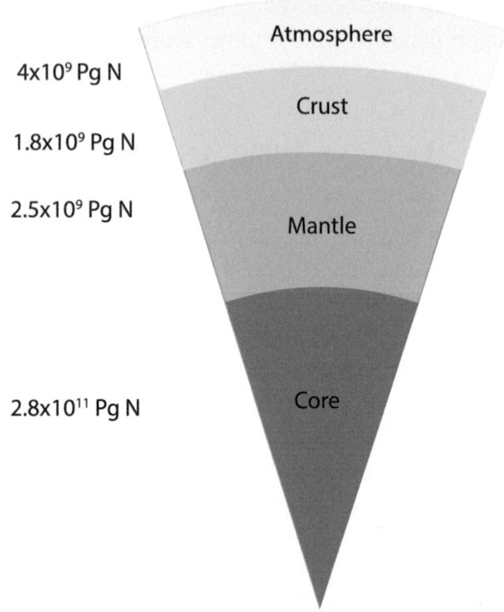

Fig. 6.7 The distribution of nitrogen in the different regions of the Earth

"faint sun paradox." A correlation between nitrogen burial and atmospheric pN2 would suggest that the biosphere could have directly driven changes in pN2 via the drawdown of nitrogen from the atmosphere, effectively forming an increasingly efficient geobiological pump of nitrogen from the atmosphere into the sediments, for subsequent storage in the deep Earth. On the other hand, an increasing pressure with time would be consistent with the tectonic style on Earth compared with planets with no such activity like Mars and Venus. At the present time, subduction zone injects oxidizing materials into the descending slab and this implies the transformation of ammonium NH_4 in N_2 with release of the latter due to it high volatility. This would also explain the higher atmospheric ratio of $N_2/^{36}Ar$ and $N_2/^{20}Ne$ with respect to the atmosphere of Mars and Venus. However considering the scarcity of data it is safe to assume that partial pressure of nitrogen has not significantly changed with respect to Archean. One interesting aspect of the nitrogen cycle is how it is represented by the different authors. We see that the only process is the nitrogen fixation by lighting that in the Archean atmosphere is enhanced by the presence of carbon dioxide emitted by the volcano according to the mechanism.

$$CO_2 + N_2 \rightarrow CO + N_2O \tag{6.58}$$

The nitrogen oxide with FeS suspended in liquid water can be reduced to ammonia NH_4^+. This compound can also be produced by nitrites (NO_2^-) and nitrates (NO_3^-) and buried in clays. The major difference with modern cycle is the absence of the biofixation due to the agriculture practices and fertilizer which amount to about 200 Tg/yr.

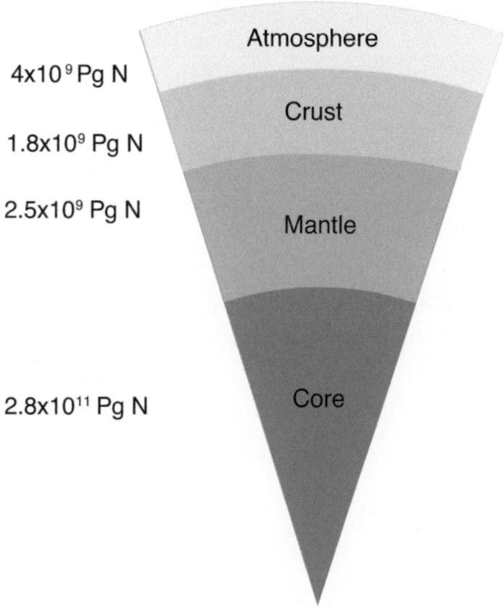

Fig. 6.8 The nitrogen cycle between atmosphere and solid Earth. Adapted from St*ü*eken et al.

Appendix

Oxidation of Methane

First of all we need to see how the hydroxil radicals are produced though the action of UV radiation

$$\begin{aligned} O_3 + h\nu &\rightarrow O_2 + O(^1D) \\ O(^1D) + M &\xrightarrow{O_2} O_3 + M \\ O(^1D) + H_2O &\rightarrow 2OH \end{aligned} \tag{6.59}$$

We have then formation of ozone and splitting of H_2O in two OH. The oxidation process of methane starts with

$$CH_4 + OH \xrightarrow{O_2} CH_3O_2 + H_2O \tag{6.60}$$

The methylperoxy radical CH_3O_2 can react with NO generating NO_2 which photolyzes producing ozone

$$\begin{aligned} CH_3O_2 + NO &\rightarrow CH_3O + NO_2 \\ NO_2 + h\nu(\lambda < 410nm) &\rightarrow NO + O \\ O + O_2 + M &\rightarrow O_3 + M \end{aligned} \tag{6.61}$$

while methylperoxy radical reacts with O_2 to give formaldehyde (CH_2O) and HO_2

$$CH_3O + O_2 \rightarrow CH_2O + HO_2 \tag{6.62}$$

Formaldehyde reacts with OH or photolyzes

$$\begin{aligned} &CH_2O + OH \xrightarrow{O_2} CO + HO_2 + H_2O \\ &CH_2O + h\nu \xrightarrow{2O_2} CO + 2HO_2 \\ &CH_2O + h\nu \rightarrow CO + H_2 \end{aligned} \tag{6.63}$$

The peroxy radical HO_2 produced by the first and second (6.63) may react with NO and generates more ozone

$$\begin{aligned} HO_2 + NO \rightarrow OH + NO_2 \\ NO_2 + h\nu \xrightarrow{O_2} O_3 + NO \end{aligned} \tag{6.64}$$

The last step is the oxidation of carbon monoxide

$$CO + OH \xrightarrow{O_2} CO_2 + HO_2 \tag{6.65}$$

Equations (6.65) and (6.64) form a catalytic cycle for ozone production catalyzed by the radicals HO_2 and NO_x with the net result

$$CO + O_2 \xrightarrow{O_2} CO_2 + O_3 \tag{6.66}$$

In the Archean atmosphere, ozone was absent so that $O(^1D)$ could not be produced. However the hydroxyl radicals resulted from reactions of hydrogen sulfide.

A Few Elementary Calculations

We make here explicit some simple calculations reported in the text. Consider 400 m of water covering the entire Earth with a surface of $5.1 \times 10^{14} \text{m}^2$. This corresponds to a volume

$$400 \times 5.1 \times 10^{14} = 2.04 \times 10^{17} m^3 = 2.04 \times 10^{23} g = 1.13 \times 10^{22} \text{mol } H_2O$$

With an escape flux of 10^{13}mol yr^{-1} the water would be dissipated in a time

$$\text{time constant} = \frac{1.13 \times 10^{22}}{10^{13}} = 1.13 \times 10^9 \text{yr}$$

1 bar of oxygen

$$\frac{5.1 \times 10^{14} \times 10^5}{9.8} = 5.2 \times 10^{21} g = 1.625 \times 10^{20} \text{ mol } O_2$$

So 40 bar of O_2 are 6.5×10^21 mol O_2 roughly equivalent to 1.13×10^{22} mol of water. This oxygen would be enough to oxidize the continents (basalts+sediments-buried carbon) in roughly 300 Myr.

Living Organism in the Ocean

We have mentioned several times photosynthesis

$$CO_2 + H_2O + h\nu \rightarrow CH_2O + O_2$$

And respiration and decay

$$CH_2O + O_2 \rightarrow CO_2 + H_2O$$

The formation of living matter is a little more complicated and must include the incorporation of elements like carbon (C), phosphorous (P), and nitrogen (N). These elements in the ocean are related by the so-called *Redfield ratio* which specifies the proportion of C,P, and N in the marine organisms. The proportion of C:N:P is 106:16:1. To represent respiration and decay in organic matter taking into account N and P photosynthesis must be represented by

$$\begin{aligned} &106CO_2 + 106H_2O + 16NH_3 + H_3PO_4 + h\nu \\ &\rightarrow (CH_2O)_{106}(NH_3)_{16}(H_3PO_4) + 106O_2 \end{aligned} \tag{6.67}$$

The opposite of (6.67) would be the respiration

$$(CH_2O)_{106}(NH_3)_{16}(H_3PO_4) + 106O_2 \rightarrow 106CO_2 + 106H_2O + 16NH_3 + H_3PO_4 \tag{6.68}$$

The terrestrial biomass is dominated by trees and then lignin so the terrestrial biomass has different proportions of C, N, P and the photosynthesis can be represented by

$$\begin{aligned} &830CO_2 + 600H_2O + 9NH_3 + H_3PO_4 + h\nu \\ &\rightleftarrows C_{830}H_{1230}O_{604}N_9P \end{aligned} \tag{6.69}$$

Now we can give some more detail about the oxygen cycle. We have connected the short-term cycle (formation of organic material through photosynthesis) to the long-term rock cycle with relations (6.6)–(6.7). There is another way and this consists in the deposition of calcium carbonate ($CaCO_3$) on the ocean bottom. $CaCO_3$ can be deposited forming calcite sediments or interact with sulfate to form gypsum sediments

$$(CaCO_3)_s + H^+ + 2H_2O + SO_4^{2-} \rightarrow (CaSO_4 \cdot 2H_2O)_s + HCO_3^- \quad (6.70)$$

The C and S sediments formed in this way remain in the lithosphere until they are brought back to the surface and weathered returning CO_2 and sulfate to the ocean. Notice that reaction (6.70) does not have any direct effect on the oxygen while both CO_2 and sulfate contents are affected and have an impact on the atmospheric oxygen.

The Modern Nitrogen Cycle

We have examined the ancient nitrogen cycle and now could be very instructive to mention the modern N cycle, especially to clarify some process. As we know nitrogen constitutes about 80% of the Earth's atmosphere as N_2. In this chemical form, nitrogen is almost inert and cannot be utilized by living organism so it must be converted through a process known as fixation to ammonia

$$N_2 + 8H^+ + 8e^- \rightarrow 2NH_3 + H_2 \quad (6.71)$$

N_2 gas is a very stable compound due to the strength of the triple bond between the nitrogen atoms, and it requires a large amount of energy to break this bond. The whole process requires 8 electrons and at least 16 ATP (Adenosine Triphosphate molecules. The energy necessary for the reaction to occurs is provided by microorganisms which use a common enzyme known as nitrogenase that catalyzes the reduction of N_2 to NH_3 (ammonia). Following the formation of ammonia is the process of *nitrification* which consists in the formation of nitrite (NO_2^-) in two steps

$$NH_3 + O_2 + 2e^- \rightarrow NH_2OH + H_2O \rightarrow NO_2^- + 5H^+ + 4e^- \quad (6.72)$$

This requires two enzymes, ammonia monooxygenase and hydroxylamine oxidoreductase. The second step in nitrification is the oxidation of nitrite (NO_2^-) to nitrate (NO_3^-). This step is carried out by a completely separate group of prokaryotes, known as nitrite-oxidizing Bacteria. There is another way to nitrificate ammonia through the so-called *Anammox (anaerobic ammonia oxidation)* carried out in anoxic condition by prokaryotes belonging to the Planctomycetes phylum of bacteria. The chemical reaction is

$$NH_4^+ + NO_2^- \rightarrow N_2 + 2H_2O \quad (6.73)$$

A quite important process is *denitrification* that is the conversion of nitrate back to nitrogen gas which is returned to the atmosphere.

$$\begin{gathered} NO_3^- \rightarrow NO_2^- \rightarrow NO + N_2O \rightarrow N_2 \\ 2NO_3^- + 10e^- + 12H^+ \rightarrow N_2 + H_2O \end{gathered} \quad (6.74)$$

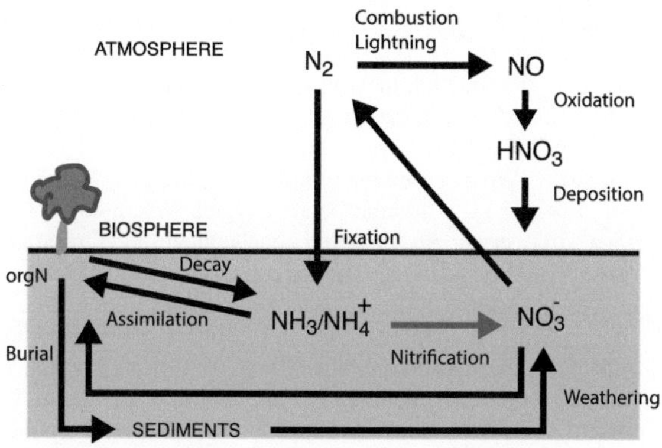

Fig. 6.9 Modern nitrogen cycle with processes discussed in the text

Denitrification produces nitrous oxide which is greenhouse gas. Biofixation that is formation of NH_3/NH_4^+ is being strongly altered by the use of fertilizer. The last process to mention is the so-called *ammonification* when the organic nitrogen (amino acid and DNA) is converted back to ammonia by fungi or prokaryotes. Figure 6.9 is another version of the biogeochemical cycle as limited to the solid Earth.

References

Canfield, D.E., Glazer, A.N., Falkowski, P.G.: The evolution and future of earth's nitrogen cycle. Science **330**, 193 (2017)

Canfield, D.E.: Oxygen: A Four Billion Year History. Princeton University Press (2014)

Canfield, D.E.: The early history of atmospheric oxygen Homage to Robert M. Garrels. Annu. Rev. Earth Planet. Sci. **33**, 1 (2005)

Catling, D.C., Claire, M.W.: How Earth's atmosphere evolved to an oxic state: a status report. Earth Planet. Sci. Lett. **237**, 1 (2003)

Catling, D.C., Kasting, J.F.: Atmospheric Evolution on Inhabited and Lifeless Worlds. Cambridge University Press (2017)

Catling, D.C.: The Great Oxidation Event Transition, in Treatise in Geochemistry, 2nd edn., vol. 6, p. 177 (2014)

Catling, D.C.: The Great Oxidation Event, Treatise on Geochemistry, 2nd ed., vol. 6, p. 177 (2014)

Claire, M.W., et al.: Biogeochemical modelling of the rise in atmospheric oxygen. Geobiology **4**, 239 (2006)

Eguchi, J., Seales, J., Dasgupta, R.: Great Oxidation and Lomagundi events linked by deep cycling and enhanced degassing of carbon. Nat. Geosci. **13**, 71 (2020)

Goldblatt, C., Lenton, T.M., Watson, A.J.: Bistability of atmospheric oxygen and the Great Oxidation. Nature **443**, 683 (2006)

Holland, H.D.: The Chemical Evolution of the Atmosphere and Oceans. Princeton University Press (1984)

Holland, H.D.: The oxygenation of the atmosphere and oceans. Phil. Trans. R. Soc. B **361**, 903 (2006)

Holland, H.D.: Volcanic gases, black smokers, and the Great Oxidation Event. Geochim. Cosmochim. Acta **66**, 3811 (2002)

Holland, H.D.: Why the atmosphere became oxygenated: a proposal. Geochim. Cosmochim. Acta **73**, 5241 (2009)
Jacob, D.J.: Introduction to Atmospheric Chemistry. Harvard University Press (2024)
Kasting, J.F.: Modeling the Archean atmosphere and climate. Treatise on Geochemistry, 2nd edn. vol. 6, p. 157 (2014)
Kasting, J.F.: What caused the rise of atmospheric O_2? Chem. Geol. **362**, 13 (2013)
Kump, L.R.: The rise of atmospheric oxygen. Nature **451**, 277 (2008)
Lasaga, A.C., et al.: An improved geochemical model of atmospheric CO_2 fluctuations over the past 100 million years. In: The Carbon Cycle and Atmospheric CO_2: Natural Variations Archean to Present. American Geophysical Union, Washington, DC, vol. 397 (1985)
Lenton, T.M.: The coupled evolution of life and atmospheric oxygen. In: Rothschild, L.J., Lister, A.M. (eds.), Evolution on Planet Earth. Ac. Press (2012)
Li, Y.: The origin and evolution of Earth's nitrogen. Nat. Sci. Rev. **11**, nwae201 (2024)
Reinhard, C.T., Planavski, N.J.: Biogeochemical Controls on the Redox Evolution of Earth's Oceans and Atmosphere. Elements **16**, 191m (2020)
Rothschild, L.J., Lister, A.M. eds.: Evolution on Planet Earth. Ac. Press (2012)
Sleep, N.H.: Dioxygen over geologic time. Met. Ions Biol. Syst. **43**, 49 (2005)
Stueken, E.E., et al.: Modeling pN2 through geological time: implications for planetary climates and atmospheric biosignatures. Astrobiology **16**, 949 (2016)
van Decker, H.K.E.: Holde. Springer, Oxygen and the Evolution of Life (2011)
Walker, J.C.G.: Evolution of the Atmosphere. Macmillan (1977)
Zahnle K.J., Catling, D.C., Claire, M.W.: The rise of oxygen and the hydrogen hourglass. Chem. Geol. **362**, 26 (2013)
Zahnle, K.J., Claire, M.W., Catling, D.C.: The loss of massindependent fractionation in sulfur due to a Paleoproterozoic collapse of atmospheric methane. Geobiology **4**, 271–283 (2006)
Zerkle, A.L., Mikhail, S.: The geobiological nitrogen cycle: from microbes to the mantle. Geobiology **15**, 343 (2017)

Chapter 7
The Climate Evolution

The atmosphere of a planet is very important for the determination of its climate. As we know well the atmospheric composition may change with time causing the climate to change on historical time scale (i.e., in a few tens of years) as it is happening at the present time due to the increasing abundance of the greenhouse gases mixing ratios. Considering the drastic changes in the composition of the atmosphere, the distribution and extension of the continents we expect to find a corresponding evolution in the climate of the Earth. There is an additional variable related to the astrophysical factor and that is that our Sun follows a luminosity evolution typical of its star class which makes it roughly 5% brighter every Ga. That makes the output of the star 30% lower when the Solar System was born. However even with such decrease in the solar output evidence shows that on our planet still had temperature which permitted the survival of life forms. This corresponds to the so-called *Faint Young Sun Paradox* that was discovered by Carl Sagan and George Mullen more than 50 years ago. First of all we will ask how we can assert the environmental conditions on such distant past and then follow possibly the evolution of the climate of the Earth until the present epoch.

7.1 How the Ancient Temperatures Are Measured (Dating Methods)

It is worthwhile at this point to define once and for all the geological time scale as shown in Fig. 7.1. We will deal with reconstruction of climate from 4.5 Ga to the present and the most important data is the temperature. We have seen in the early chapters that the temperature of the Earth in all the Archean should have been quite hot because of the heavy asteroid bombardment, and the residual heat of formation including the impact which formed the Moon. Now we ask ourselves whether there are proofs of such events and this is a very hard question to answer because the

G. Visconti, *Origin and Evolution of the Atmosphere and Ocean*,
https://doi.org/10.1007/978-3-032-02653-8_7

Fig. 7.1 The chronology of the different eras on Earth

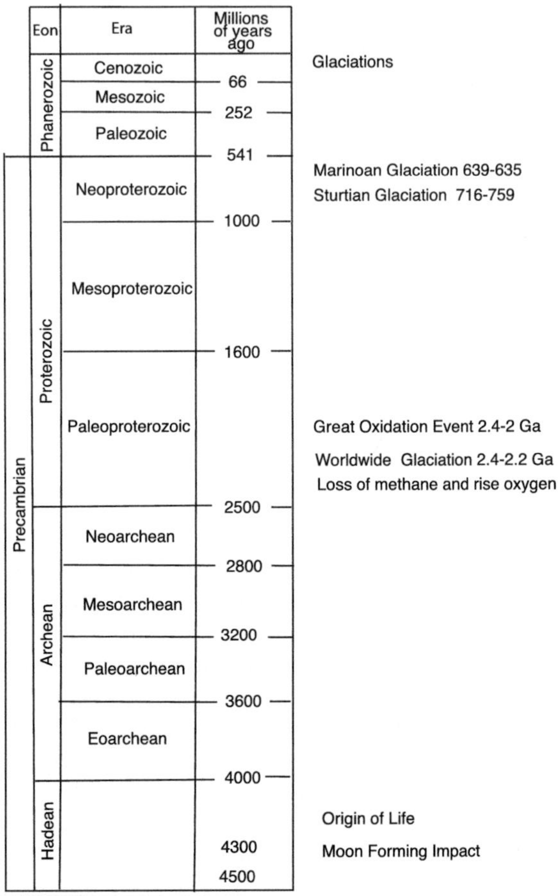

evidence is rather sparse. Until a few years ago it was thought the climate of the Earth was *hell-like* and the name Hadean was reflecting such idea. Just before the turn of the twenty-first century John Valley and coworkers (2002) were able to study in detail the nature of *zircon minerals* $(ZrSiO_4)$ which are included in the oldest known rocks (4.4 Ga). The age of these minerals was determined by the uranium ìmpurities in the crystal and for an Australian rock it was found to be 4.4 Ga. Then they measured the isotope ratio $^{18}O/^{16}O$ to determine the temperature of formation (see Chap. 5). In the mantle such ratio is on the average ≈5.3‰ while they found values up to 7.4‰. The only explanation they could find was that such minerals had interacted with rain and ocean water. Water from rain is enriched in the heavier isotope. The conclusion was that after the formation of the Moon the planet cooled much more rapidly to the point of permitting liquid water to exist on the surface. Similar conclusion based on the so-called *zircon thermometer* was reached by Watson and Harrison a few years later. For the remaining periods different methods are used to determine the temperature.

We already know about the $\delta^{18}O$. Since ocean temperatures are, relatively, stable the $\delta^{18}O$ value mostly reflects the temperature where precipitation occurs. The relation with temperature can be simplified as

$$\delta^{18}O = aT + b \tag{7.1}$$

where a varies between 0.33 and 0.76‰/°C. A similar approach is used for carbon using the fact that during photosynthesis ^{12}C is taken preferentially with respect to ^{13}C because it is lighter so that organic matter has a lower content of ^{13}C with respect to the atmosphere. The evaluation of

$$\delta^{13}C = [(^{13}C/^{12}C)_{sample}/(^{13}C/^{12}C)_{standard} - 1] \times 1000$$

can be made on both the bubbles contained in the ice cores and in carbonate sediments (see the discussion of Lomagundi Event on Chap. 6). Another thermometer is based on the ratio Mg/Ca that is magnesium to calcium in planktic/indexplanktic (surface waters) and benthic (deep waters) foraminifera. The incorporation of magnesium impurities depends on temperature in the sense that this ratio increases with temperature. The same happens for impurities of strontium (Sr). More recent methods are based on the records of organic products like alkenones which are long-chain unsaturated methyl and ethyl n-ketones produced by a few phytoplankton species.

7.2 A Brief History of Climate on Earth

Before going for the different topic could be important to review briefly the climate behavior during the age of the Earth. Naturally the data are scarce and contradictory for the distant past and become more and more detailed up to the Quaternary, which is a period within the Cenozoìc.
a. *Precambrian Climate*. This covers the time between the origin of the Earth up to about 540 Ma ago and include the *Hadean*, (up to 4Ga) the *Archean* (up to 2.5 Ga) and the *Proterozoic* (up to 541 Ma) Fig. 7.1. Data for this period seem to confirm the presence of an ocean and this constitutes one of the elements of the so-called *Faint Young Sun (FYS) paradox*. At that time the luminosity of the Sun could have been 30% lower than today and this would imply freezing temperature on Earth which apparently was not the case according to various kinds of geological evidence. The paradox was ascertained by Carl Sagan and George Mullen in a famous *Science* paper more than 50 years ago and its explanation has remained fundamentally the same (an enhanced greenhouse effect) as we will see later. As we have seen this epoch was characterized by the Great Oxidation Event (GOE) that signed the initial raise of oxygen in the atmosphere and the birth of the ozone layer. Both of these events brought about and acceleration of methane destruction and a reduction of greenhouse effect. The distribution of the continental area was quite different during the Precambrian and before its end the continents were grouped in a single mass

mostly around the present south pole corresponding to the super-continent *Rodinia* (Donnadieu et al. 2004). This may have influenced the climate of this time which was characterized by a number of glaciations. The most severe were around 2.4–2.2 Ga (roughly coincident with the GOE and the abatement of methane) and 759–635 Ma (the Sturtian and Marinoan glaciations, respectively). The last one may have coincided with the *Snowball Earth* a period in which all the globe was covered with ice.

b. *Phanerozoic climate* This covers the eon from about 540 Ma to the present epoch and includes the *Paleozoic* (up to 250 Ma) the *Mesozoic* (up to 68 Ma) and the *Cenozoic*. As we have seen the Earth's climate has remained quite stable and the only notable deviations are represented by the glaciation episode. So even in the Phanerozoic we have three main episodes: a Late Ordovician/early Silurian (455–425 Ma), a late Devonian and brief episode ca 360 Ma, and a Permo-Carboniferous glaciation between 335 and 260 Ma. This was followed by a warming period with considerable temperature culminating about 100 Ma. Then we had a sudden collapse of temperature in correspondence of the late Cenozoic Glaciation about 35 Ma. The Mesozoic era between 250 and 65 Ma includes the Triassic, Jurassic, and Cretacious periods and was characterized by a large transfer of carbon from solid Earth to the atmosphere and the ocean which produced a very warm climate. The Mesozoic ended with the impact of the *Dinosaur Asteroid* that was responsible for the disappearance of the large reptiles. As we have mentioned the end of Cretacious was followed by the Cenozoic glaciation. The climate during this time era probably influenced by the separation of South America from Antarctica. This cooling was then followed by a Miocene warming and finally approaching the Quaternary a huge buildup of polar caps and the glaciation episodes that were cycling with periods that were initially of 40 kyr (from 3 Ma to about 1.5Ma) and then changed to 100 kyr. The history of the last glaciation ($\approx$20 kya) is very well documented. So the relevant climatic phenomena we will discuss should include the *Faint Sun Paradox, the Snow Ball Earth, the Dinosaurs extinction, the Glaciations*. Figure 7.2 inspired to the work of (Zalasiewicz and Williams 2009) is an efficient summary of the Earth's climate history since its origin.

7.3 The Faint Sun Paradox

The Faint Sun Paradox (FSP) was proposed in a paper published in 1972 and it is based on the fact that stars of the Sun class (a G star) increase the luminosity with time. This implies that our star was much dimmer 4.6 Gya. We can follow the old paper by Michael Newmann and Richard Rood that for the luminosity

$$L \sim \frac{ac}{k_0} M^{5.5} R^{-0.5} \left(\frac{G \mu M_H}{k} \right)^{7.5} \tag{7.2}$$

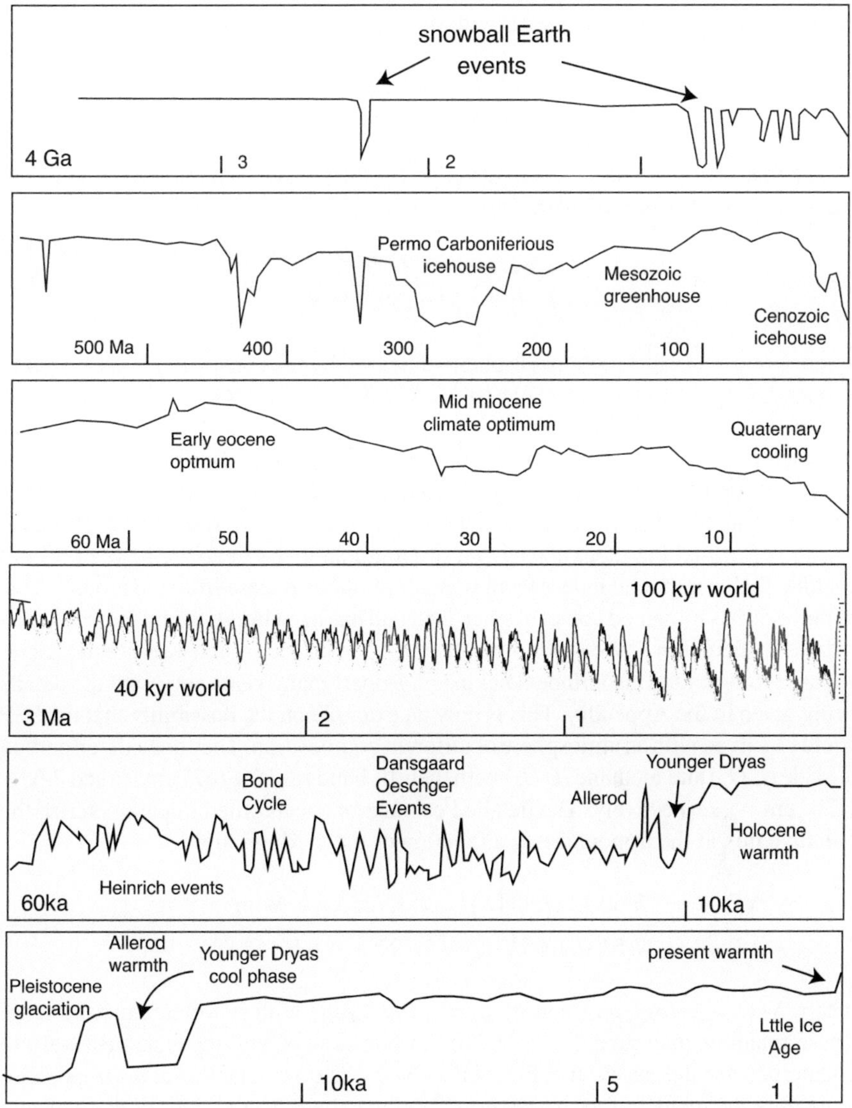

Fig. 7.2 Qualitative behavior of the Earth temperature for different eras. Adapted from Zalasiewicz et al. (2020)

where a is the radiation density constant $4\sigma/c$ with σ Stefan–Boltzmann constant, k the Boltzmann constant, M and R mass and radius of the star, M_H the hydrogen mass, G the gravitational constant, and μ the average molecular weight. k_0 some kind of absorption coefficient. All the quantities appearing in (7.2) remain constant except for the molecular weight μ which increases because of the hydrogen burning.

As a consequence the change of L with time is given by

$$\frac{1}{L}\frac{dL}{dt} \sim \frac{7.5}{\mu}\frac{d\mu}{dt} \tag{7.3}$$

The change with time of μ can be related to the mean hydrogen mass fraction X and the Z the mass fraction of elements heavier than helium so that (7.3) becomes

$$\frac{1}{L}\frac{dL}{dt} \sim \frac{12.5L}{M\epsilon(1+1.66X)} \tag{7.4}$$

where $\epsilon \approx 6.4 \times 10^{18} erg/g$ is the energy released by the conversion of hydrogen to helium. Using the values $M = 2 \times 10^{33} g$, $X = 0.75$, $L = 3.84 \times 10^{33} erg\ sec^{-1}$ we obtain $1/\tau = (1/L)dL/dt \approx 0.05/10^9 years$ and the change with time of the solar constant $F_s = F_0 \exp(t/\tau)$ with $F_0 \approx .75 * 1360\,\text{watt m}^{-2}$. The same inventors of this paradox proposed that an enhanced greenhouse effect produced by an appropriate amount of methane (CH_4), ammonia (NH_3), and carbon dioxide (CO_2). However it was shown that the rapid conversion of ammonia to N_2 in about 40 years would exclude such compound unless there was a continuous release from the ground. This hypothesis has remained constant since then and has been improved by more detailed radiative calculations. We report here the ones performed by Jeff Kiehl/indexKiehl, Jeff and Robert Dickinson/indexDickinson, Robert many years ago, with the details being given in the Appendix. This is only an example on the possibility that the FSP could be solved with an atmosphere of different composition. They considered carbon dioxide (CO_2) and methane (CH_4) with their IR bands at 15 µ (677 cm^{-1}) and 7.65 µ (1307 cm^{-1}), respectively. The detailed calculations were fitted to parameterize the radiative flux at the tropopause as a function of the mixing ratio

$$\begin{aligned} -\Delta F_{CO_2} &= 73.60 - 27.434X_{CO_2} + 1.725X^2_{CO_2}\ \text{W m}^{-2} \\ -\Delta F_{CH_4} &= 27.53 - 1.65X_{CH_4} - 1.39X^2_{CH_4} + 0.158X^3_{CH_4}\ \text{W m}^{-2} \end{aligned} \tag{7.5}$$

where $X_{CO_2} = -log(\mu_{CO_2}$ and $X_{CH_4} = -log(\mu_{CH_4})$ with μ_i volume mixing ratio. These relations may give the amount of carbon dioxide and methane necessary to counterbalance the variation of the solar constant. The net solar flux at the tropopause is roughly the incident flux (≈340 w m^{-2}) minus the reflected flux. With an albedo around 0.3 we have as net flux $\approx S_n = 238$ w m^{-2} so ΔS_n, the change in IR flux necessary to compensate the change in solar flux, is given by

$$\frac{\Delta S_n}{S_n} = f_s = 1 - \frac{S}{S_0} \tag{7.6}$$

which gives

$$-\Delta F_{CO_2} - \Delta F_{CH_4} = \Delta S_n = f_s S_n \tag{7.7}$$

where S_0 is the reference solar constant and S the is the value at some time. Substituting (7.5) we have

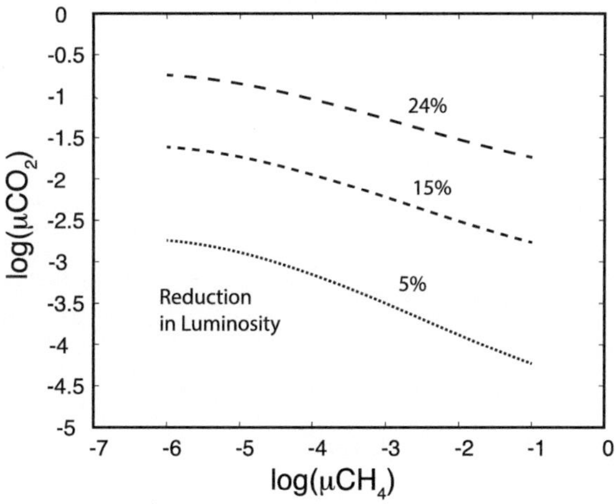

Fig. 7.3 Mixing ratios of methane and carbon dioxide necessary to compensate for the reduction in solar luminosity

$$X_{CO_2} = 7.952 - (5.647 + 130.42 f_s + 0.957 X_{CH_4} + 0.806 X_{CH_4}^2 - 0.0916 X_{CH_4}^3)^{1/2} \tag{7.8}$$

This expression is plotted in Fig. 7.3 where we see that for a given reduction of the solar flux the amount of carbon dioxide required to compensate the effect is a function of the methane mixing ratio. However the mixing ratios of both gases results to be in the range of 100–200 times the present values.

In a paper published in 2010, Minik Rosing/indexRosing, Minik and coworkers examined the constraints to the mixing ratios of the two gases given by the geologic evidence. They examined the process through which methanogens bacteria digest the hydrogen

$$CO_2 + 4H_2 \rightarrow CH_4 + 2H_2O \tag{7.9}$$

The rate of this reaction depends on temperature and the mixing ratios of carbon dioxide and hydrogen. Methanogens could reduce the partial pressure of H_2 down to $10^{-5} bar$ with a CO_2 pressure ranging between $10^{-3.5}$ and $10^{-4.5} bar$. These values are compatible with the deposits of magnetite ($FeOFe_2O_3$) and siderite ($FeCO_3$). The conclusion is an Archean atmosphere with an H_2 partial pressure of $10^{-5} bar$ and a CO_2 mixing ratio about three times the present value. Such amount of greenhouse gases would not be enough to compensate for a faint sun. In the same paper, they gave an alternative explanation for the paradox based on the reduction of the albedo of the planet. This idea was based on the fact that in the early history of the Earth the continental distribution was quite different from the present one. The first thing to observe is the difference between the *planetary albedo* (α_{planet}) and the *mean surface albedo* ($\alpha_{surface}$). The first which is now measured with space-borne instruments depends on the first but also include contributions from the atmosphere and the clouds. The α_{planet} is given by

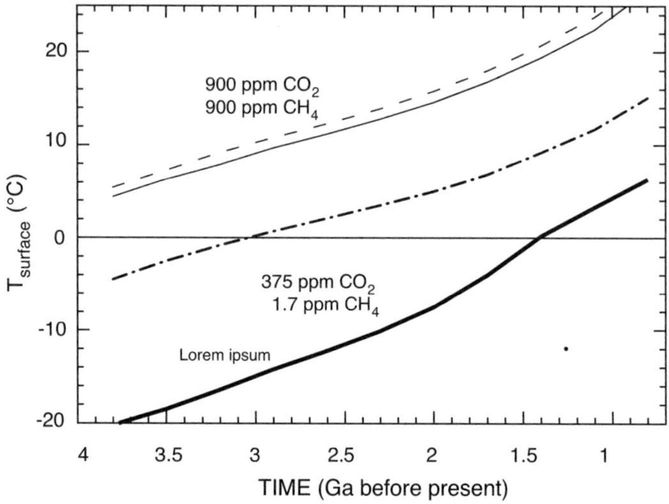

Fig. 7.4 Temperature of the Earth as a function of time for different amounts of greenhouse gases and albedo (adapted from Rosing et al.)

$$\alpha_{planet} = \eta_{land}\alpha_{planetland} + \eta_{ocean}\alpha_{planetocean} \tag{7.10}$$

where $\eta_{land} = 0.3$ is the present area fraction of the atmosphere underlain by land, and the ocean fraction is $1 - \eta_{land}$. At the present time we can assume $\alpha_{planetland} = 0.35$ and $\alpha_{planetocean} = 0.3$ so that $\alpha_{planet} = 0.315$. On the other hand, the mean surface albedo can be calculated from the contributions of the different surfaces. The ocean has an average albedo $\alpha_{surfocean} \approx 0.1$ while the land $\alpha_{surfland} \approx 0.23$ so that

$$\alpha_{surface} = 0.23 \times 0.3 + 0.1 \times 0.7 = 0.14 \tag{7.11}$$

But this can be further specialized to include different components, for example, ice covering the ocean or land so that

$$\begin{aligned} \alpha_{surface} = \alpha_{surfocean}[\eta_{ocean} - (1 - \eta_{oceanice}]+ \\ \alpha_{surfland}[\eta_{land} - (1 - \eta_{land\ ice}]+ \\ \alpha_{surface\ ice}(\eta_{ocean}\eta_{ocean\ ice} - \eta_{land}\eta_{land\ ice}) \end{aligned} \tag{7.12}$$

Expression (7.12) can be used to evaluate the change of surface albedo with time. So the author assumed that η_{land} would change from about 5% to the present 30%. On the other hand, $\eta_{land\ ice}$,$\eta_{ocean\ ice}$ and $\alpha_{ice}.\alpha_{ocean}$ are as today. Then $\alpha_{surface}$ would grow from 0.08 to 0.15 about 1 Gya to decrease later to 0.13 with the appearance of the biosphere. These values were increased when calculating the planetary albedo essentially for the presence of clouds characterized by the droplet size.

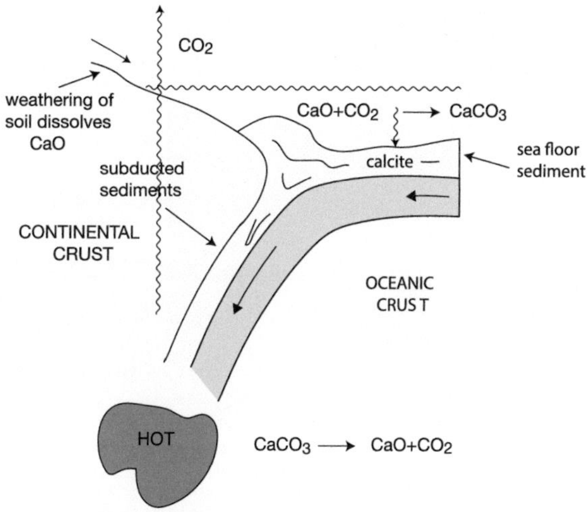

Fig. 7.5 The simplified cycle of carbon dioxide in the lithosphere

The results show values from 0.22 to about the present values around 0.3. With such values of the albedo and different mixing ratios for CO_2 and CH_4 then results are schematically shown in Fig. 7.4. It obvious that large mixing ratio ($\approx$900 ppm) and lower albedo could maintain reasonable temperatures.

7.4 The Long-Term Stabilization of Climate

Before going to examine the snowball Earth puzzle it is worth to examine the proposal for the long stabilization of climate on the planet. This problem is somewhat reduced to the long-term control of carbon dioxide and the hypothesis was put forward by James Walker and coworkers in a very simple but illuminating paper in 1981. Wally Broecker simplified further the scheme as shown in Fig. 7.5. The crust is assumed to be composed only of CaO which is dissolved by weathering and brought to the sea. Here it combines with the dissolved CO_2 to form calcium carbonate $CaCO_3$

$$CaO + CO_2 \rightarrow CaCO_3 \tag{7.13}$$

The formed calcite is deposited on the seafloor and transported by plate tectonics in the deep hot mantle where it is again decomposed in CO_2 and $CaCO$ completing the cycle. Carbon dioxide is released from the interior and if this rate is larger than the $CaCO$ supply it accumulates in the atmosphere raising the temperature. This causes an increase in the weathering and so the supply of $CaCO$ to the sea reducing the amount of CO_2 in the atmosphere. Actually a more accurate reaction involves wollanstonite mineral ($CaSiO_3$) which again consumes CO_2

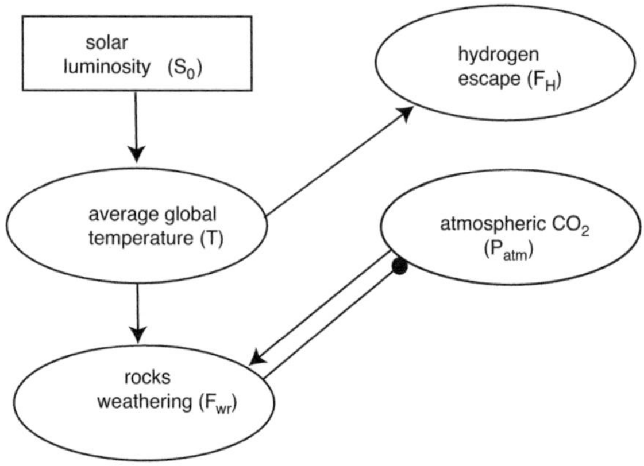

Fig. 7.6 The interactions between different elements of the Earth. Arrows denote the direct influence while dots are for opposite sign

$$CaSiO_{3_s} + CO_{2_g} \rightarrow SiO_{2_s} + CaCO_{3_s} \tag{7.14}$$

The links which relate the different quantities are shown in Fig. 7.6 and now to lay down the equations we will follow the original 1981 just to illustrate the simplicity of the scheme. There is a subsequent paper by Ken Caldeira and James Kasting which discusses also the effects on the biosphere and we will talk about it in one of the next paragraphs. The first step, however, is to parameterize some of the processes in very simple terms. We can start from the solar radiation, assuming it increases with time according to the data we have on stellar evolution with a relation similar to (7.6)

$$S(t) = \frac{S_0}{(1 - 0.38t/\tau_0)} \tag{7.15}$$

where S_0 is the present solar constant (1368 $w\ m^{-2}$), $\tau_0 = 4.55$ Ga while t is the time expressed in Ga and is negative. At the birth of the Earth this expression gives a value for S of about 72% of the present value. The crucial assumption in this model is to establish how the weathering changes with temperature. The data in this case are quite uncertainty and as a little academic exercise to assume that the weathering rate W depends on the partial CO_2 pressure p and the emission rate from volcano V

$$\frac{W}{W_0} = \left(\frac{p}{p_0}\right)^{0.3} \exp\frac{\Delta T}{13.7} \tag{7.16}$$

where "0" refers to the present values. When we equate (7.16) to the emission rate, we get

$$\frac{V}{V_0} = \left(\frac{p}{p_0}\right)^{0.3} \exp\frac{\Delta T}{13.7} \rightarrow \frac{p}{p_0} = \left(\frac{V}{V_0}\right)^{10/3} \exp -\frac{\Delta T}{4.11} \tag{7.17}$$

If $T_{:e}$ is the effective temperature determined by the solar flux the surface temperature T_s will be increased by the greenhouse effect due to the carbon dioxide and water

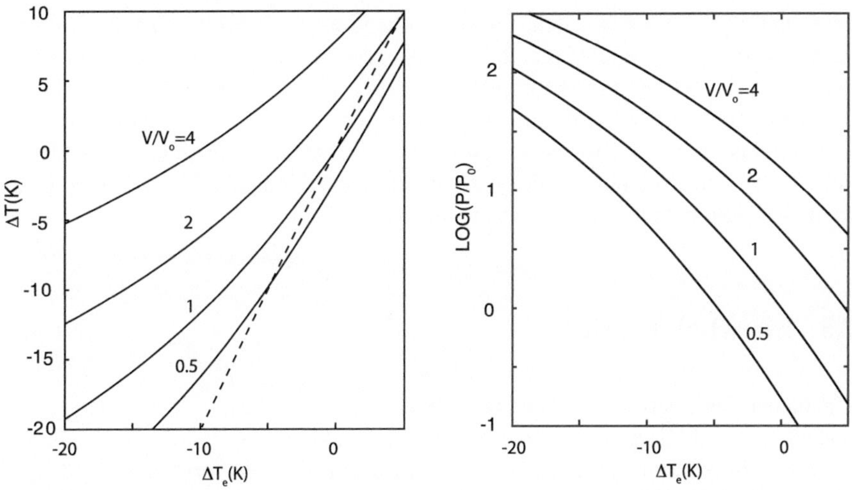

Fig. 7.7 On the left the increment in temperature ΔT as a function of the effective temperature ΔT_e for different emission rates V/V_0. The dashed line is for constant pressure. On the right the CO_2 pressure as a function of ΔT_e for different emission rates

vapor

$$T_s = T_e + T_g(CO_2) + T_g(H_2O) \tag{7.18}$$

These increments were taken from existing models at that time and are given by

$$T_g(CO_2) = 2.3\left(\frac{p}{p_0}\right)^{0.364} \quad , \qquad T_g(H_2O) = 29.7 + 0.5(T_s - 285) \tag{7.19}$$

Then we can evaluate (7.18) for the present conditions ($p = p_0, T_s = 285, T_e = 253$) and subtract it from the generic condition to obtain

$$\Delta T = 2\Delta T_e + 4.6\left[\left(\frac{p}{p_0}\right)^{0.364} - 1\right] \tag{7.20}$$

where $\Delta T = T_s - 285$. $\Delta T_e = T_e - 253$. In Fig. 7.7 we show the relations between ΔT and ΔT_e and between (p/p_0) and ΔT_e for different values of the emission rate V/V_0. Also shown is the relation when the feedback is absent ($p/p_0 = 1$) and the emission rate is constant ($V/V_0 = 1$). In this case, we have from (7.17) and (7.20)

$$\frac{p}{p_0} = \exp -\frac{\Delta T}{4.11} \quad , \quad \Delta T = 2\Delta T_e$$

As we would expect partial pressure decreases with increasing ΔT. Negative values of ΔT_e correspond to lower values of the present solar flux. The authors use value of the solar flux as was known at the time of publication (1327 watt m^{-2}) with an albedo of 0.3 to get $T_e = 253$ K. If this flux is reduced by 25% we would get $T_e = 235$ K that is $\Delta T_e = -18$ K and the upper part of Fig. 7.7 would give $\Delta T = -18$ K so that the temperature of the Earth would still be too cold $T = 267K$ with $V/V_0 = 1$. The conclusion is that this mechanism is not enough to compensate for the lower luminosity of the Sun but the idea is smart and we will talk about in a different way when will deal with the survival of the biosphere.

7.5 Snowball Earth

The planet Earth between 720 and 600 Mya could have suffered an extended glaciation which would have covered the entire planet making it very similar to the Jupiter satellite Europa. The hypothesis known *Snowball Earth* was formulated by Joseph Kirschvink in 1992 to explain the presence of glacial deposits at tropical latitudes.

The theoretical basis to explain the snowball Earth was laid down by Mikhail Budyko and William Sellers which developed in the late 60s of the last century the so-called Energy Balance Models (EBM) with the introduction of the *Ice–Albedo Feedback*. The idea is that the extension of the ice caps is a function of the temperature: if for some reason the temperature decreases the extension of the ice caps increases and so does the albedo which increases further the cap's extension. This positive feedback may result in an ice catastrophe with the planet entirely covered with ice. The details of such model are given in the *Appendix* in the simple two-mode version proposed by Gerry North and the results are summarized in Fig. 7.8. This figure is redrawn from the work of Caldeira and Kasting published on *Nature* in 1992. The idea was that once the ice catastrophe you deal now with a sphere with quite high albedo (0.62) so that an elementary calculation would give a quite large increase in

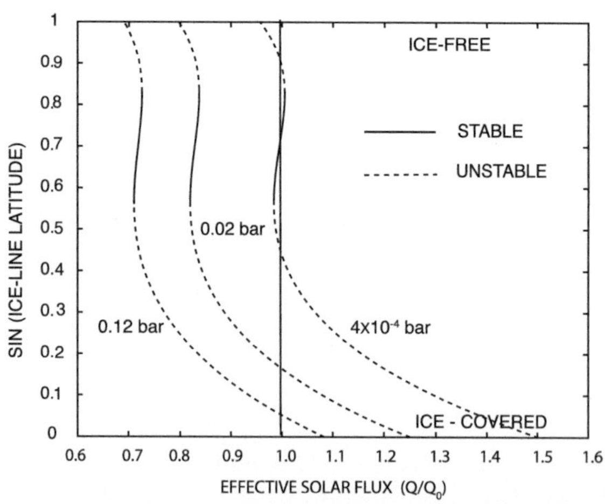

Fig. 7.8 The sine of the ice line latitude as a function of the carbon dioxide pressure and the solar constant. The dashed lines are regions of instability

the solar constant to exit the snowball situation

$$\frac{q}{4} = \frac{\sigma T^4}{(1-\alpha)} \rightarrow \frac{q}{4} \approx \frac{5.67 \times 10^{-8} \times 273^4}{(1-0.68)} \approx 984\text{wm}^{-2} \approx 2.9 \times q_0 \quad (7.21)$$

There is another solution to the problem which has been proposed and that is the volcanic activity that would pump huge quantities of carbon dioxide in the atmosphere. A calculation can be performed with the Energy Balance Model and the results (details in the appendix) discussed in the same figure. Notice that the intersection of the line $q/q_0 = 1$ results in three equilibrium points only one of which correspond to stable climate (see Fig. 9.11 for a complete discussion) while the upper one can be assumed to be the latitude limits for the pole cap. Notice that increasing the carbon dioxide the beginning of the instability leg moves toward lower values of the q/q_0 ratio and the same happens for the zero latitude for the ice cap until the curve hits the $q/q_0 = 1$ mark. At this point the snow cover is unstable in a situation of constant q. This simple thought experiment was first proposed by ken Caldeira and James Kasting and they found that the critical value for the carbon dioxide mixing ratio was 0.2 bar of partial pressure while we have found about 0.183 bar. Many other hypotheses have been formulated for the snowball (including the ones that deny it) but we hope to have given an idea of the complexity of the problem.

7.6 The Extinction of Dinosaurs

In 1980, the nuclear physicist Luis Alvarez (and coworkers) published a paper on *Science* claiming that the dinosaurs' extinction dated about 66 million years ago was caused by a giant impact of an asteroid with diameter around 10 km and an energy between 0.1 and 2.4 billion of megaton (1 Megaton $\approx 4.18 \times 10^{15}$ J). After a long search the impact point was confirmed to be on the coast of Yucatan Peninsula and it known as Chicxulub Crater. One of the proofs brought by Alvarez et al. was that a geologic layer corresponding to that epoch (the limit Cretaceous-Paleogene, K-Pg boundary) contained anomalous amount of iridium. The same metal was found in the Chicxulub Crater. The asteroid impact is not the only theory about dinosaur extinction the other most popular attribute is the disappearance to huge episodes of volcanism which deprived the dinosaur of their preferred environment. The theory about the effect of the asteroid impact was formulated initially by Owen Toon and coworkers in 1982 in a paper of the Geological Society of America and on *Science* a year later. The same group of people about a year later revisited another problem raised earlier by Paul Crutzen and John Birks which revisited an idea in the early 70s related to the environmental effects of a nuclear war. The group was referred to as TTAPS (from the author initials, Turco, Toon, Ackermann, Pollack, and Sagan). These effects were quite similar to those of an asteroid impact and could be grouped in the Chemical and Climatic consequences. For the latter, we will follow a quite simple treatment given by Arthur Broyles. Let us finish with a simple calculation of the asteroid mass.

Measurements give an average iridium concentration of about $3 \times 10^{-7}\,\mathrm{Kg\,m^{-2}}$. So that the total mass on Earth of iridium is the surface density multiplied by the total surface $3 \times 10^{-7} \times 5.1 \times 10^{14} = 1.53 \times 10^{8}\,\mathrm{kg}$. The average concentration of iridium in carbonaceous chondrites is about 5×10^{-7} and consequently the mass of the asteroid could be $\approx 1.53 \times 10^{8}/5 \times 10^{-7} \approx 3 \times 10^{14}\,\mathrm{kg}$. With a bulk density of $2.2 \times 10^{3}\,\mathrm{kg\,m^{-3}}$ this gives a radius of $\approx$3.2 km. So we assume an object of 10 km diameter which has a kinetic energy at the impact $\approx 0.5 \times 3 \times 10^{1} \times (2.5 \times 10^{4})^{2} \approx 9.6 \times 10^{22}\,J \approx 2 \times 10^{7}$ Megaton of the same order of the previous data.

7.6.1 Climatic Effects

We start with the consideration that the impact of the asteroid produced not only a huge dust cloud (Morgan et al. 2022) but also widespread fires due to the high temperatures developed during the event (Santa Catharina et al. 2022) which devastated 70% of the existing forest on the planet. The problem could be then compared to that of the nuclear winter as examined by Broyles in a rather simple way. He assumes that the fires produce smoke and the cloud occupies an area between 30 and 70 latitude which corresponds to about $A_c = 1.1 \times 10^{14}\,\mathrm{km}^2$. We may refer to the latitude of Chicxulub Crater which is about 20° and consider an area comprised between 0 and 40° of about $1.6 \times 10^{14}\,\mathrm{km}^2$. The optical thickness of the cloud is roughly

$$\tau = tn\sigma_e \tag{7.22}$$

where t is the thickness of the cloud in meter, n is the density of particles, and σ_e is the extinction cross section. There is a way to relate the number of particles and thickness to the yield of the bombs Y and the emission efficiency of smoke (E). Also it can be shown that once we have fixed the average radius of particles (0.1 µm) the optical thickness varies with the inverse of the horizontal area A_c and we assume

$$\tau \approx 0.27 Q_e E Y \tag{7.23}$$

with Q_e extinction efficiency $Q_e = \sigma_e/2\pi r^2$. Another missing parameter for our calculation is the single scattering albedo (ω) of the particles. Broyles lists some of this relevant data that we have summarized in Table 7.1.

With these parameters it is now possible using a two-stream approximation (see Appendix) compute how much radiation reaches the ground. The transmission, T, and the reflectivity, R, are given by

$$T = \frac{\alpha}{\alpha \cosh \alpha\tau + a \sinh \alpha\tau} \qquad R = \frac{b \sinh \alpha\tau}{\alpha \cosh \alpha\tau + a \sinh \alpha\tau} \tag{7.24}$$

where τ is the optical thickness, $a = 1 - \omega f$, $b = \omega(1 - f)$, $\alpha^2 = a^2 - b^2$ with ω single scattering albedo and f is the fraction of backscattered radiation. For very

Table 7.1 Possible smoke data for a 1000 Mt war. Table is partly reproduced from Broyles. The optical thickness has been reduced because the smoke cloud now extends between the equator and 40°

E			0.001	0.63
	Q_e	ω	τ	τ
Smoldering wood	0.41	0.9986	0.11	6.8
Flaming wood	0.68	0.74	0.18	11.5
Carbon	2.19	0.42	0.58	36.7

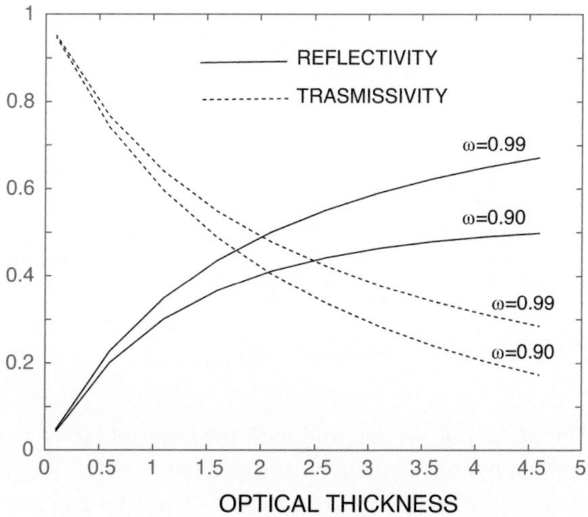

Fig. 7.9 Transmissivity, T, and reflectivity, R, as a function of optical thickness and single scattering albedo ω

large values of τ we have $R = b/(\alpha + a)$, $T = 0$. Figure 7.9 gives the behavior of T and R and T for several values of the optical thickness. Now we follow Broyles in evaluating the effect of smoke on the surface temperature. Notice that this approach is mainly based on the difference between the behavior of pressure and density with height. Indicating the altitude h in km we have for pressure p and density ρ

$$p = p_0 e^{-0.13\,h}, \qquad \rho = \rho_0 e^{-0.104\,h} \tag{7.25}$$

This slight difference is derived from the fitting to real data when the temperature also changes with height. Then we notice that energy E in a diatomic perfect gas is given by

$$E = \frac{5}{2} NkT \tag{7.26}$$

where N is the total number of molecules, k is the Boltzmann constant, and T is the temperature. If n is the number of mole then $n = N/N_a$ and $k = R/N_a$ and (7.2) becomes

$$U = \frac{E}{V} = \frac{5}{2}p \tag{7.27}$$

So energy density is proportional to pressure. If the smoke raises to an altitude h_s the energy per unit area is given by

$$\int_0^{h_s} U dh = \frac{5}{2}\int_0^{h_s} \rho dhk\langle T\rangle = \frac{5}{2}\int_0^{h_s} pdh \tag{7.28}$$

Extracting $\langle T\rangle$ and performing the integrations, we have

$$\langle T\rangle = \frac{0.104(1 - e^{-0.13h_s})T_0}{0.13(1 - e^{-0.104h_s})} \tag{7.29}$$

Assuming an altitude $h_s \approx 9km$ and a temperature $T_0 = 288K$ we obtain $\langle T\rangle \approx 261K$ that is a cooling of about 27 K. This exercise refers to a total yield of 1000 Megaton while the asteroid was several magnitudes larger. The conclusion is level of darkness quite severe that lasted several years.

7.6.2 *Atmospheric Composition*

The impact of the asteroid with the atmosphere of the Earth has the main effect of producing huge amount of nitrogen oxide (NO). When the major constituents of the air, nitrogen, and oxygen are heated to high temperature, nitric oxide (NO) is formed.

$$N_2 + O_2 \leftrightarrow 2NO \tag{7.30}$$

As the shock-heated air cools, the NO mixing ratios decrease until we reach the quench temperature where the time constant for NO removal exceeds the time constant for cooling. The yield of NO has been estimated between 0.5×10^{16} molecules of NO/Joule up to 30×10^{16} molecules of NO/Joule and the quench temperature changes between 2000 K for nuclear explosion to 2700 K for lightning. The production of NO can be approximated by

$$P = \tfrac{1}{2}mv^2[\epsilon_1 Y_1 + (1 - \epsilon_1)\epsilon_2 Y_2] \tag{7.31}$$

where ϵ_1 is the fraction of the incoming bolide kinetic energy deposited during entry into the atmosphere and Y_1 is the corresponding NO yield, ϵ_2 is the fraction of the bolide kinetic energy at surface impact which is returned to the atmosphere as kinetic energy in the high-velocity ejecta plume and Y_2 is the corresponding NO yield, m is the bolide mass, and v is the bolide speed. The value of ϵ_1 is quite small ($<10^{-3}$) with ϵ_2 much bigger (≈ 0.125). The produced NO in the entry plume has a mixing ratio of the order of 10^{-4} (notice that in the present conditions NO mixing ratio is around 300 ppb) and with these concentrations NO is converted in nitrogen dioxide through the reaction

$$NO + NO + O_2 \rightarrow 2NO_2 \tag{7.32}$$

This thermochemical reaction essentially is irreversible since the rapid ultraviolet photodissociation of NO_2 which occurs in the present-day troposphere is precluded through ultraviolet shielding by the overlying high-altitude NO_2. Most of the nitrogen dioxide formed in this way will produce dinitrogen oxide (N_2O_4)

$$NO_2 + NO_2 + M \rightarrow N_2O_4 + M \tag{7.33}$$

This can be followed by hydration to form nitrous (HNO_2) or nitric acid (HNO_3).

$$H_2O + N_2O_4 \rightarrow HNO_2 + HNO_3 \tag{7.34}$$

Another way to form acid is through the passage to nitrogen pentoxide (N_2O_5). The acid rain formed has a pH around 1 (the present-day pH is around 4, see Appendix) while the NO_x mixing ratio would have reached 10^{-4} with the present values around 3×10^{-10}. For the mentioned NO mixing ratio, the daylight levels are extremely low and the photosynthesis ceases to exist. The predicted acid rain leads to defoliation of plants and to asphyxiation of lung-respiring animals. The effect of defoliation and cessation of photosynthesis will severely deplete the food supply for vegetarians; one would expect the largest animals to be affected the greatest. As it is well known once the acid precipitation reaches the ground a major effect is the mobilization of metals which has several and dangerous effects.

Clearly the large NO_x mixing ratio after the impact has also the effect to deplete the ozone through the catalytic cycle

$$\begin{aligned} NO + O_3 &\rightarrow NO_2 + O_2 \\ NO_2 + O &\rightarrow NO + O_2 \\ net, \quad O_3 + O &\rightarrow 2O_2 \end{aligned} \tag{7.35}$$

Other consequences of an asteroid impact must be considered and have to do with the ocean. The most interesting one arises from a possible impact with the ocean. This possibility has been studied in detail some years ago by Elisabetta Pierazzo. The study dealt with relatively small size asteroid (up to 1.5 km) but it is interesting because of its estimate of NO production which amounts to

$$m_{NO} \approx 10^{-3} m_{H_2O} \tag{7.36}$$

However as mentioned by Birks et al. an oceanic impact could result in the production of halogens compounds that could destroy ozone. The chemical reactions involve salt particles (NaCl) and chloride, bromide, and iodide ions contained in the sub-micron-sized salt particles containing chloride, bromide, and iodide ions

$$
\begin{aligned}
ClONO_2 + NaCl(s) &\rightarrow Cl_2 + NaNO_3(s) \\
N_2O_5 + NaCl(s) &\rightarrow ClNO_2 + NaNO_3(s) \\
HNO_3 + NaCl(s)) &\rightarrow HCl + NaNO_3(s)
\end{aligned} \tag{7.37}
$$

Free chlorine could result and trigger ozone destruction through the well-known catalytic cycle.

$$
\begin{aligned}
Cl + O_3 &\rightarrow ClO + O_2 \\
ClO + O &\rightarrow Cl + O_2 \\
O_3 + h\nu &\rightarrow O + O_2 \\
net\ \ 2O_3 &\rightarrow 3O_2
\end{aligned} \tag{7.38}
$$

Some of the huge amount of water vapor produced by the impact could photodissociate producing OH and H that could contribute to the ozone destruction through the odd hydrogen catalytic cycle.

7.6.3 Ocean Acidification

The consequences of impact on the ocean composition were theorized many years ago by John Lewis and now we have new and more detailed theories on that. In the appendix, we explain that the present acidity of the ocean is determined by the total dissolved inorganic carbon (DIC)

$$
[DIC] = [CO_2(aq)] + [H_2CO_3] + [HCO_3^-] + [CO_3^{2-}] \tag{7.39}
$$

Increase in [DIC] correspond to lowering the pH that at the present time is around 8.5. A first sign of the Ocean Acidification (OA) after the impact is the extinction of many pelagic calcifying species (including 100% of ammonites and more than 90% of calcareous nanoplankton and foraminifera). It is estimated that the impact vaporized about 6×!0^{16} mol of sulfur from gypsum rocks and 5×10^{15} mol of NO_x and also the release of up to 6,500 Pg C as CO_2 from vaporization of carbonate rocks, wildfires, and soil carbon decay. Finally ocean overturn brought high CO_2 water to the surface. We would like to resume some numbers about the content of carbon dioxide in the atmosphere. With a mixing ratio of 400 ppm corresponding to a mass mixing ratio of $4 \times 10^{-4} \times 44/29 \approx 6 \times 10^{-4}$ and the atmospheric mass of 5.2×10^{21} g we arrive to total mass of CO_2 roughly 3.12×10^{18} g. A pulse of 6500 Gt of C correspond to 5.4×10^{17} mole of carbon dioxide with a mass of 2.30×10^{19} g that is about 7 PAL. This estimation reported by Tyrrel et al. is based on the total burnt vegetation and the carbon in the soil that was not replenished by litter during the extended period of darkness. The asteroid itself could have contained gypsum which is a calcium sulfate dihydrate ($CaSO_4 \cdot 2H_2O$) so that its impact could have produced sulfur trioxide according to

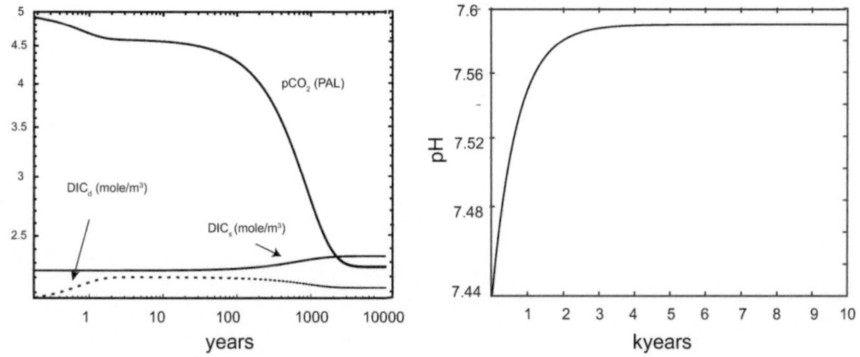

Fig. 7.10 The response to a carbon dioxide pulse (left) and the consequent pH behavior (right)

$$CaSO_4 \rightarrow CaO + SO_3 \tag{7.40}$$

Once ejected in the atmosphere SO_3 would form sulfuric acid (H_2SO_4) in a process well known from the volcanic eruptions. Subsequently this acid would be rained out contributing to the ocean acidification. In the same paper, it is estimated a contribution up to 1900 Pg of sulfur (SO_4). Finally production of NO_x could have contributed to the formation of roughly 5×10^{15} g of nitric acid (HNO_3). We have performed a simulation for carbon dioxide with the model reported in the appendix for a pulse equivalent to raising the carbon dioxide content up to 5 PAL and the results are shown in Fig. 7.10 together with examples taken from Tyrrel et al. It can be seen a basic difference between addition of CO_2 on one side and HNO_3 and H_2SO_4 on the others. In the first case, acidification happens progressively while the direct deposition of acid particle decreases instantly the pH value with recover in a few years.

7.7 The Ice Ages

We all know that Earth suffered, with some regularity, periods of cold which extended the polar caps to mid-latitude. The last one terminated about 20.000 years ago while the previous ones interested the last million years with a period of about 10^5 year. We can only give a short summary of the problem which has countless sides and characteristics. We may refer mainly to the paper by Maureen Raymo and Peter Huybers and the interesting contribution of Didier Paillard. Raymo and Huybers published the beautiful figure similar to Fig. 7.11 which depicts the behavior of the δ^{18}O(‰). In the original figure also the obliquity and the summer insulation and finally the solstice insulation are compared. Only the isotope record, however, is produced from experimental data and we notice that the oscillations around 1 million years ago changed their period from about 40,000 year to 100,000 year showing a

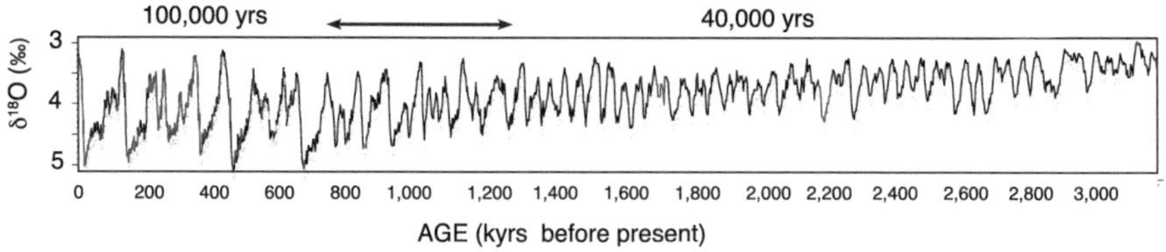

Fig. 7.11 The δ^{18}O(‰) for the last 3 Mya. To notice the change of the oscillation period from roughly 100 kya to 40 kya

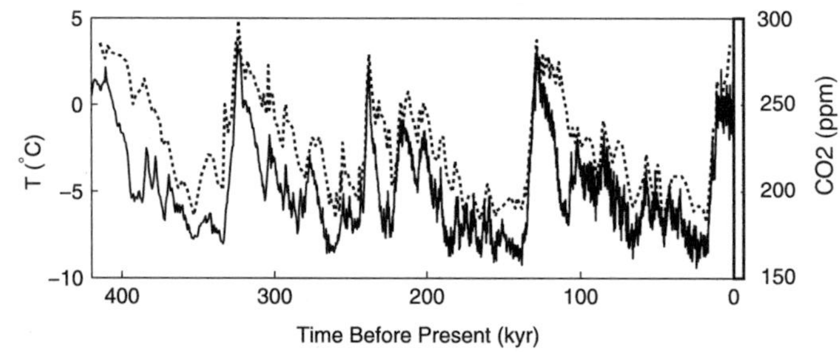

Fig. 7.12 The temperature record and the CO_2 pressure (dashed lines) for the last five glaciations

progressive asymmetry between the buildup of ice caps (in about 90 kyr) and their decline (in about 10 kyr). The correlation between the ice volume and the obliquity is quite evident while things are more complex if related to insolation at the solstice. In this figure, unfold about the glaciations are summarized the main question. Why the period changed from the prevalent 41 to 100 kyr and why there is the pronounced asymmetry in the ice buildup and demise. Actually the only experimental data in the figure refer to the δ^{18}O(‰) which as we have seen in Chap. 5 can be related to the temperature or better to the ice volume. There is another piece of evidence which adds another enigma to ice age problems. It the early 70s, the composition of the air contained in the bubbles trapped in the ice cores revealed the record of carbon dioxide that as shown in Fig. 7.12 follows closely the temperature (ice volume) behavior. Is the classical chicken and egg problem. In what follows, we will try to explain the different aspects related to the glaciations.

7.7.1 *The Astronomical Forcing*

We will start considering some characteristics of the Earth orbit as summarized in Fig. 7.13. a. *Eccentricity* The orbit is an ellipse and the eccentricity e is defined as the ratio between the focal distance c and the semi-major axis a, $e = c/a$ and as consequence the semi-minor axis is $b = a(1 - e^2)^{1/2}$. The eccentricity changes with time with periods of $\approx 10^5$ and $\approx 4 \times 10^5$ years. The present value is 0.0167.

Eccentricity bears a direct effect on the orbit of the Earth which is described by the conservation of angular momentum

$$r^2 \frac{d\nu}{dt} = \frac{2\pi ab}{T} = \frac{2\pi a^2}{T} \sqrt{(1-e^2)} \tag{7.41}$$

where r is the distance from the Sun, ν is the frequency, and T is the period. When mediated over a complete revolution we have the relation between the solar constant S_0 and the power received S in wm^{-2}

$$S_0 = \frac{S}{T} \int_0^T \frac{a^2}{r^2} dt = S \int_0^{2\pi} \frac{d\nu}{2\pi \sqrt{1-e^2}} = \frac{S}{\sqrt{1-e^2}} \tag{7.42}$$

Considering that the value of e varies between 0 and 0.06 we have $S_0 \approx (1 + e^2/2)S \approx (1 + 0.0018)S$, that is, the increase is only of the 0.18%. The change of the power received is comparable with the variation of the solar output.

b. *Obliquity*

Obliquity ϵ represents the tilt of the rotation axis of the Earth with respect to the ecliptic plane and because of the gyroscopic motion the associated equinox precession. The present value of ϵ is about 23.5° while actually oscillates between 21.9° and 24.5° with a period of 41.000 years. Changes in obliquity do not have effects on the total radiation received by the Earth but rather influences its geographical distribution. A larger value of ϵ implies that polar regions receive more radiation while the tropics less. It can be shown that the solar power received in a year at the pole and the equator results as

$$\begin{aligned} W_{year}(pole) &= \frac{S}{\pi\sqrt{1-e^2}} \sin\epsilon \qquad wm^{-2} \\ W_{year}(equator) &= \frac{2S}{\pi^2\sqrt{(1-e^2)}} E(\sin\epsilon) \qquad wm^{-2} \end{aligned} \tag{7.43}$$

where $E(sin\epsilon) = E(\pi/2, \sin\epsilon)$ is the complete elliptic integral of the second kind. These relations give for e = 0, a difference of 1% for the $W_{year}(pole)$ (≈18 wm^{-2}) and 0.4% for $W_{year}(equator)$ (≈5wm^{-2}) for the excursion of obliquity. A similar change exists for the solstice but just for the $W_{summer}(pole) = S \sin\epsilon$ because in such case $W_{winter}(pole) = 0$. Then change in the received radiation in this case is between $S \sin 21.9$ and $S \sin 24.5$ that is of the order of 50 wm^{-2}.

c. *Precession*

The gravitational action of the Sun and Moon on the Earth's equatorial bulge is responsible of the precession of its axis. This has a period of 25,765 years and the principal effect is a drift of the equinoxes around the orbit of the planet. This movement, however, must be combined with the *precession of perihelion* that for the Earth has a period of about 112,000 years. This precession is known for Mercury and

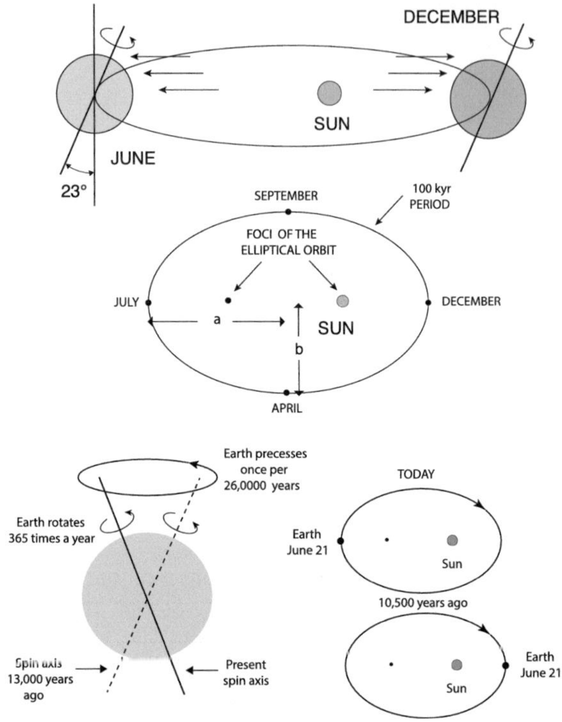

Fig. 7.13 The geometry of the astronomical forcing

connected with one of the proofs of the theory of general relativity. For the Earth, the precession of perihelion has opposite sign with respect to the equinox so that the result in period is given by

$$\frac{1}{25.7} + \frac{1}{112} \approx \frac{1}{21}$$

At the present time, at the summer solstice, the Earth is near its aphelion and in about 10,500 years the summer solstice will coincide with the perihelion producing an evident modulation in the solar radiation received. Notice that the distance from the Sun changes between a(1+e) and a(1-e). The combination of the eccentricity frequency ($\approx 1/100$) with the equinox frequency ($\approx 1/21$) gives raise to two more frequencies

$$\frac{1}{21} + \frac{1}{200} \approx \frac{1}{19}, \qquad \frac{1}{21} - \frac{1}{200} \approx \frac{1}{23}$$

Periods of 23,000 and 19,000 years were actually detected in the ice records. This phenomenon is quite similar to the generation of tones when the sum of two waves with quite equal frequencies ω_1 and ω_2 give rise so to low-frequency $\Delta\omega = \omega_1 - \omega_2$ and I high-frequency carrier wave.

7.8 How to Explain Glaciations

As of today there is not a simulation of the onset of a glaciation using, for example, a General Circulation Model or an Earth System Model because we can't afford the cost of modeling even a single glaciation cycle (100,000 years). All we got are then ideas or toy models which reproduce some aspects of the ice ages. The main problem of the ice age is the fact that in the last four million years the period has changed from the obliquity to the eccentricity and apparently the astronomical forcing related to the latter is not enough to cause any build up of the polar caps. This episode is known as Mid-Pleistocene Transition (MPT). It is now quite accepted (see Willeit1 et al. 2019) that such a transition is due to the fact that in the last 3 million years the carbon dioxide mixing ratio has decreased progressively from about 350 ppm to about 250 ppm. This has prolonged the build up of ice sheets that become even more stable because of the disappearance of their regolith foundations. As for the change from 41 to 100 kya one of the proposal is that once the accumulation of ice starts it continues until its mass is such to force the Earth crust to deform in such way that the ice sheets move to warmer regions starting the process of melting (Watts and Hayder 1983). In this case, the 100.000 year period is related to the characteristics of the solid Earth. Another class of theories is based on the hypothesis that the climate has multiple states and glaciations represent the transition from different states (Paillard 1998). In this class of theories we could also comprehend the stochastic resonance when the noise energy (the weather) is tuned to the weak eccentricity signal. Finally the last class of theories invokes some kind of frequency modulation operated by the weaker forcing (Rial 1999). In what follows, we outline some of these theories with details given in the appendix.

7.8.1 The 100 Kyr Cycle and the Lithosphere

This idea was originated by Johannes Weertman and was resumed by Robert Watts, and Ehteshamul Hayder. The model used in this case is sketched in Fig. 7.14. The

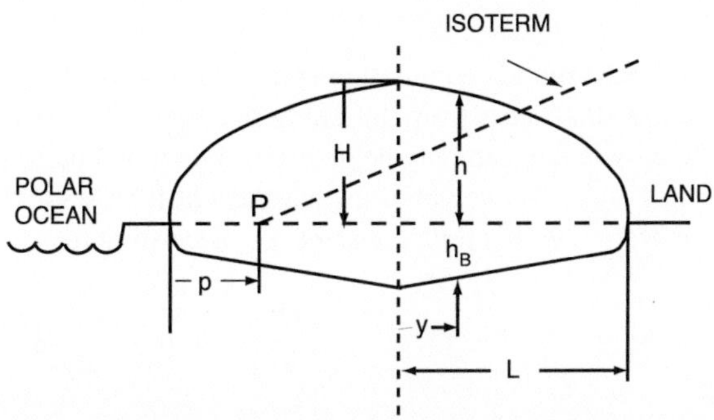

Fig. 7.14 The ice sheet

shape of the ice sheet remains fixed as it grows and adapts as a perfect plastic material. The shape is fixed with a relation between the height, h, of the ice sheet and its length, L, according to

$$h = \sqrt{\lambda(l - y)} \tag{7.44}$$

so that the maximum thickness H is given by $H = \sqrt{\lambda l}$. The quantity h_B is the buried height and 2L the latitudinal extensions. The other quantities relevant to the problem must be introduced in connection with the isotherm shown in the figure. This isotherm actually is the snow line: above this line we will assume an accumulation of ice, while below we will have ablation. The north in this figure is on the left so that going to the left we will have lower temperatures. Considering that the temperature decreases also with altitude, it is clear that the snow line will slope upward on the right side. Once the slope is fixed, the position of this line can be obtained by fixing the distance between the intersection (P) of the snow line and the surface to the north. We would expect that if the global temperature decreases, the point P will move to the south so that the surface available for the growth of the ice sheet will increase. The portion of the ice sheet above and to the north of the snow line is denoted as accumulation zone and the portion to the south ablation zone. Other than the mentioned contribution by Weertman, these models are also due to Hans Oerlemans and more recently to Cornelius J. Van der Veen. In 1983, R. G. Watts and E. Hayder added to the ice model a simple energy balance model. From observations (and also from a bit of intuition), the rate of growth for the ice sheet can be written as

$$\begin{aligned} A =& aL[H] - s[2L - p] - bL[H] - s[2L - p]^2, \quad \text{if} \quad H - s(2L - p) < a/2b \\ A =& \frac{a^2 L}{4b}, \qquad\qquad \text{if} \quad H - s(2L - p) > a/2b \end{aligned} \tag{7.45}$$

The growth rate of the ice sheet is actually the rate of increase of the ice sheet (assumed to have a unit width) cross section. Equation (7.45) needs some comment. The quantity $H - s(2L - p)$ is the difference between the top of the ice sheet and the ice line measured on the southern edge. The net ice accumulation increases with this difference until we have

$$H - s(2L - p) = \frac{a}{2b}$$

This maximum accumulation rate corresponds to the southernmost position of the ice line (that starts from the southern edge of the ice sheet) so that beyond this point the growth rate can only be given by the second part of Eq. (7.45). The Earth's crust bends under the weight of the ice sheet as it grows and that the same crust adjusts to this weight with a time constant α^{-1} according to

$$\frac{dh_B}{dt} = \alpha \left(\frac{h}{3 - h_B} \right) \tag{7.46}$$

So that at the steady state we have $h_B = h/3$ and we also assume $h_B/h = \eta(t)$ is only function of time. So that at the steady state we have $h_B = h/3$ and we also assume $h_B/h = \eta(t)$ is only function of time. The rate of change of the volume for a unit width of the part of the ice sheet that results toward the equator is

$$\frac{d}{dt}\int_0^L (h + h_B)dy = \frac{2}{3}\sqrt{\lambda}\frac{d}{dt}[(1+\eta)L^{3/2}] \tag{7.47}$$

Now we normalize the variables as follows:

$$R^2 = \frac{4s^2L}{\lambda}, \quad p = \frac{2s^2P}{\lambda} \quad \tau = \frac{at}{2}, \quad \kappa = \frac{b\lambda}{2as}, \quad \mu = \frac{2\alpha}{a}, \quad \Gamma = \eta R \tag{7.48}$$

So that combining (7.45) and (7.47) we have

$$\begin{aligned} \frac{1}{R^2}\frac{d}{d\tau}\left[\left(1+\frac{\Gamma}{R}\right)R^3\right] &= (R - R^2 + p)[1 - \kappa(R - R^2 + p)] \\ \frac{1}{R^2}\frac{d}{d\tau}\left[\left(1+\frac{\Gamma}{R}\right)R^3\right] &= \frac{1}{2\kappa} \end{aligned} \tag{7.49}$$

The first being valid for $(R - R^2 + p) < 0.5\kappa$ and the second for $(R - R^2 + p) > 0.5\kappa$. Also we have

$$\frac{1}{\mu}\frac{d\Gamma}{d\tau} = \frac{R}{3} - \Gamma \tag{7.50}$$

We also require $R > 0$. Now we can verify that the model works by supposing that the position of the climate point P changes periodically with the law

$$P = P_o + \Delta P \sin\left(\frac{2\pi t}{T}\right) \tag{7.51}$$

This forcing can be put into non-dimensional form using

$$\beta_0 = \frac{2s^2P_0}{\lambda}, \qquad \beta_1 = \frac{2s^2\Delta P}{\lambda}, \qquad \omega = \frac{4\pi}{aT} \tag{7.52}$$

So that (7.51) becomes

$$p = \beta_0 + \beta_1 \sin\omega\tau \tag{7.53}$$

To integrate (7.49) and (7.50) we use $s \approx 10^{-3}$, $\lambda \approx -7\,\text{m}$, $P_0 \approx -150\,\text{km}$, $\Delta P \approx 300\,\text{km}$ $a \approx 0.73 \times 10^{-3}\,\text{yr}^{-1}$, $b \approx 0.27 \times 10^{-6}\text{m}^{-1}\text{yr}^{-1}$, $T = 20,000\,years$, $\alpha^{-1} \approx 5000\,years$ so that we have $\beta_0 = -0.0429$, $\beta_1 = 0.0714$, $\kappa = 1.29$, $\omega = 0.938$, $\mu = 0.55$. Regarding these numbers we can make some considerations. First of all, the slope of the isotherm is simply the ratio between the temperature lapse rate and its horizontal gradient and can be easily determined. In this way, the

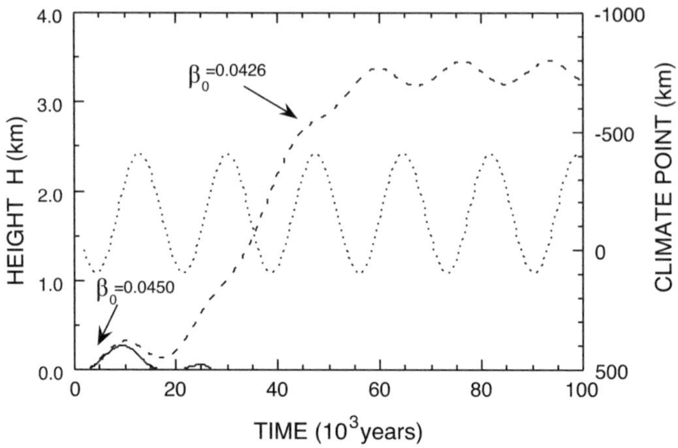

Fig. 7.15 The height of ice sheet as a function of time. The dotted line is the climate point

maximum growth is for a/2b = 1350 m. To obtain the real time from the normalized one, we need to multiply by 2740 years. We may ask with all the approximations we have made if it makes any sense to write down numbers to the fourth decimal digit but we will discover the reason for this in a while. Figure 7.15 shows the results of the integration. In the figure are shown as a function of time, the maximum height of the glacier for two different values of the forcing function that corresponds to a change in the middle position of the climate point of only 8.4 km. We notice clearly a bifurcation because a small change in P_0 corresponds to the growth or the decay of the ice sheet. The initial growth is rather slow and the modulation due to the solar radiation (that in this case has a period of 20,000 years) is quite evident. Further insight into the mechanism can be obtained by building a diagram in the phase space defined by the dimension of the glacier and the climate point, as shown in Fig. 7.16. Starting from an initial value $p(\tau)$ zero as the climate point moves toward south (that is, the Earth cools), the glacier will grow. Once the maximum value of $p(\tau)$ is reached, the growth slows down even when p becomes negative. If p stays negative long enough, the ice sheet will melt completely and in order to start to grow again p must become positive again. In this way, we enter a limit cycle and the ice sheet grows but not to a large size. In the case that p becomes positive before the ice sheet is completely melted the size will increase even when p again becomes negative. Then the ice sheet will grow to a maximum size and a new limit cycle will begin. This bifurcation does not depend only on the value of β_0, which corresponds to the initial position of the climate point, but also on β_1 which corresponds to the amplitude of the oscillation. To understand this point we may consider the case in which the crust can adapt instantaneously to the ice sheet weight ($\mu = \infty$).

We can then study the system near its equilibrium point for small values of R. In this case (7.10) could be linearized and gives

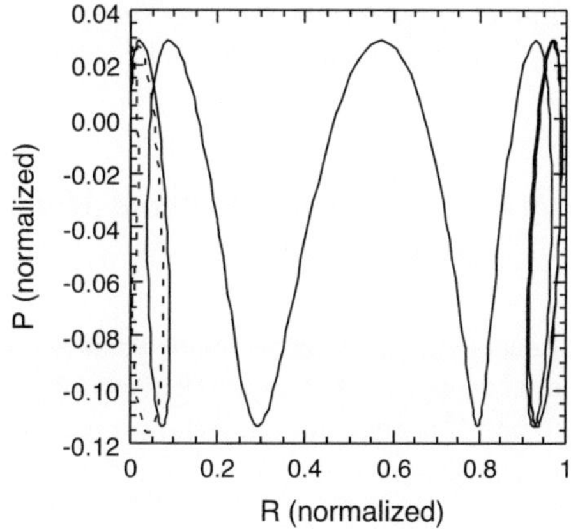

Fig. 7.16 The phase space diagram for the growth of the ice sheet. On the axis is the position of climate point p and the dimension R. Both are normalized

$$\frac{4}{3}\frac{dR}{d\tau} = R + p \tag{7.54}$$

With initial condition $R = 0$ when $p = 0$, i.e., when $\tau_1 = \omega^{-1} \arcsin(-\beta_0/\beta_1)$ and again when $\tau_2 = \omega^{-1} + \tau_1$ it results that

$$-\frac{\beta_0}{\beta_1} = \frac{1}{[1 + (4\omega/3)^2]^{1/2}} \tag{7.55}$$

As a consequence, if β_0/β_1 is larger than the right-hand side, then the ice sheet will remain small. On the other hand, if β_0/β_1 is smaller, then $R(\tau_2) > 0$ and the ice sheet will grow reaching a limit cycle around R = 1. To understand how this second limit cycle works we put $R = 1 + \epsilon$, with ϵ a small quantity, so that we can write a linearized form for Eq. (7.49)

$$\frac{4}{3}\frac{d\epsilon}{d\tau} = p - \epsilon \tag{7.56}$$

with solution

$$\epsilon = \beta_0 + \frac{\beta_1}{[1 + (4\omega/3)^2]^{1/2}} \sin[\omega\tau - \arctan(4\omega/3)] \tag{7.57}$$

As a consequence, the center of this limit cycle is at $R = 1 + \epsilon \approx 1 + \beta_0 \approx 1$ that is $L \approx \lambda/4s^2 \approx 1.7 \times 10^3\ km$. The theory then says that in order for the ice sheet to grow the cooling is not enough (i.e., β_0). This could, for example, explain why until 2.4 billion years ago there is no trace of glaciation on the Earth, because the climate was too warm. When further cooling produced a colder climate, a threshold was reached for the bifurcation to occur and consequently the ice sheets grew. This theory does not explain the sudden termination of the ice ages. However, a variation of the same theory, from David Pollard, suggests a mechanism for the sudden warming.

In practice, when the ice sheet starts to melt on its southern edge, it may form a glacial lake that may accelerate its ablation or calving.

7.8.2 A Multi-state Climate Model for the Timing of Glaciations

Michel Ghil has proposed many years ago a simple idea that could justify the oscillation of the climatic system between the glacial and interglacial. He observed that the rate of change of temperature is related to the albedo by

$$\frac{dT}{dt} \propto -\text{albedo} \propto -V_{ice}$$

With V_{ice} the ice volume. In order to close the system we need to know how this ice volume evolve and a possibility is to relate it to net the precipitation p that is

$$\frac{dV_{ice}}{dt} \propto p$$

where $p \approx p_{ac} - p_{ab}$ that is the difference between accumulation and ablation. If we now assume further that the net precipitation is related to temperature, $p \propto T$ we have

$$\frac{d^2T}{dt^2} \propto -\frac{dV_{ice}}{dt} \propto -T \tag{7.58}$$

which has an oscillatory behavior.

The model proposed by Didier Paillard in 1998 is a typical example of multi-state climate behavior. We have mentioned the possibility of multiple state of climate in connection with the snowball Earth and the idea at the base of the model is that the climate system as illustrated in Fig. 7.17 as three possible states, the interglacial **i**,

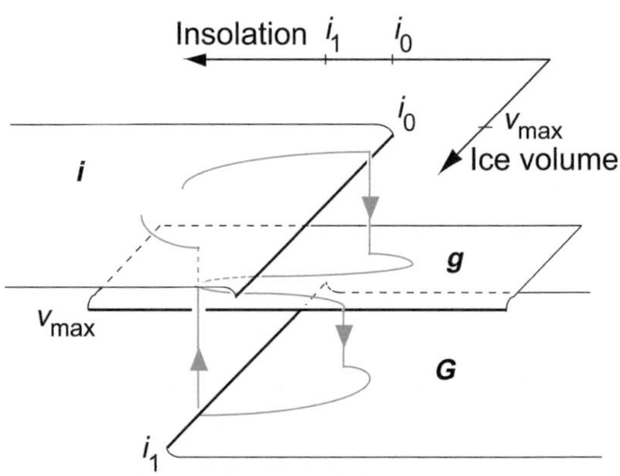

Fig. 7.17 The stages of the Paillard model

the full glacial **G**, and the mild glacial **g**. The glaciations starts (**i** → **g**) when the insolation falls below some level i_0. It proceeds to full glaciation (**g** → **G**) when the ice volume reaches some value V_{max} and finally to deglaciation (**G** → **i**) when insolation increases to some value i_1. The equation determining the ice volume evolution at each "mode" or "regime" (indicated by the index R which can take the values **i**,**g** , **G**) is

$$\frac{dV_{ice}}{dt} = -\frac{V_R - V}{\tau_R} - \frac{F}{\tau_F} \tag{7.59}$$

where V_R is the ice volume in one of the three regimes, V is the current volume, F is the forcing, and τ_R and τ_F are time constants. The volume is normalized to unity so that

$$V_g = V_G = V_{max} = 1, \quad V_i = 0 \tag{7.60}$$

The equation for the ice volume is solved numerically in a very simple way. Using the index ì+1 for the successive time steps, the difference equation becomes

$$V(i+1) - V(i) = \frac{V_g(i) - V(i)}{\tau_R} - \frac{F}{\tau_F} \tag{7.61}$$

So that the volume V(i) at the previous time is

$$V(i) = [V(i+1) + V_R(i)/\tau_R - F(i)/\tau_F]\tau_R/(1+\tau_R) \tag{7.62}$$

In the appendix, we list the MATLAB program which evaluates both the insolation and the ice volume. Figure 7.18 shows some results of this model.

7.8.3 Stochastic Resonance

The mechanism of stochastic resonance was proposed by a group of Italian physicists in the early 1980s. They proposed a theory to explain the 100.000-year cycle of the ice age. This period coincides with a change in the eccentricity of the Earth's orbit which in principle could modulate the solar radiation reaching the Earth. However, such changes are too small to have any effect and the reasoning was that when this forcing signal was summed to the noise (the weather) on some occasion the glaciation could be triggered. The mechanism is illustrated in Fig. 7.19. A sinusoidal signal is summed with the noise. When this composite signal reaches some threshold a pulse is produced. It is clear that the probability of such events higher when the sinusoidal signal has a positive value so that the pulse spectrum should have the frequency of the periodic signal. The stochastic resonance could not explain the origin of the $\approx 10^5$ years periodicity but it has found so many other applications that is worth mentioning. Consider then the energy balance equation

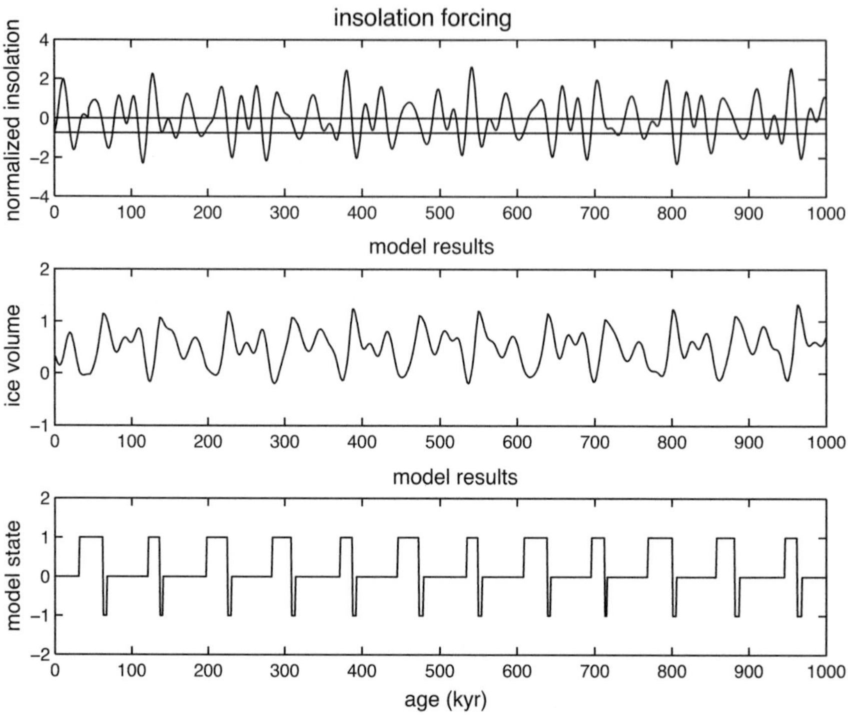

Fig. 7.18 A typical result of the program simulating the ice ages. The figure at the bottom gives the most simple models where the system is simply triggered among the three states, i, g, and G

$$\frac{dT}{dt} = R_{in}(T) - R_{out}(T) + \sigma\eta(t) = F(T) + \sigma\eta(t) = -\frac{\partial\Phi}{\partial T} + \sigma\eta(t) \tag{7.63}$$

where $R_{in}(T)$ and $R_{out}(T)$ are the incoming and outgoing radiation terms while and $\sigma\eta(t)$ is a white noise term due to internal noise of the climate system (e.g., weather). While it is obvious the temperature dependence of the second term the incoming radiation depends on temperature through the albedo. The potential function $\Phi(T)$ is such that it has two minima where $\partial\Phi\partial T$ at the equilibrium points T_1 and T_2. These are separated by an unstable temperature T_3. The potential function could be written as

$$F(T) = -\frac{\partial\Phi}{\partial T} = \frac{\epsilon(T)}{C} \frac{\mu(t)}{1 + \beta(1 - T/T_1)(1 - T/T_2)(1 - T/T_3)} \tag{7.64}$$

The eccentricity forcing is represented by $\mu(t) = 1 + 5 \times 10^{-4} \cos(\omega t)$ with $\omega = 2\pi/10^5 years^{-1}$. The effect of $\mu(t)$ is to lower and raise the two minima of the potential with respect to the middle barrier. There are possible two cases where the white noise is not large enough to extract the system from one of the minimum wells

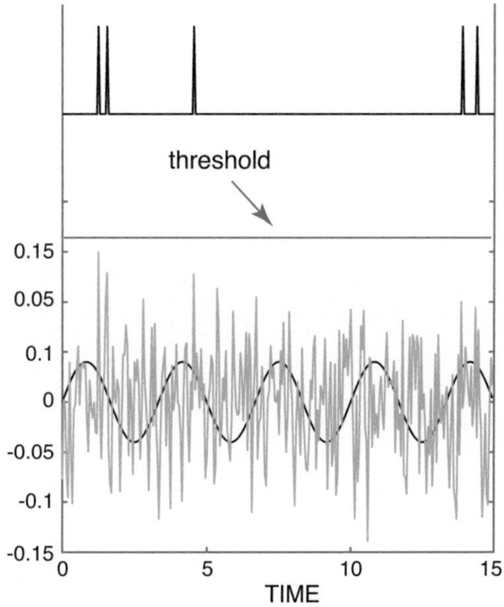

Fig. 7.19 The mechanism of stochastic resonance. A periodic signal is summed up to noise. A signal is triggered each time a threshold will be hit

or it is such to allow a transition between T_1 and T_2. Because this mechanism relies on the 100 kyr Milankovitch signal which is very weak climatically, and because the resulting time series does not have the saw tooth structure, the stochastic resonance is not a likely candidate for the 100 kyr glacial cycle mechanism.

7.9 Glacial Cycles and Carbon Dioxide

One of the most remarkable facts emerging from the analysis of the core drilling is the evident correlation between the temperature and the CO_2 abundance as it is shown in Fig. 7.20. The conundrum of the *hen and egg problem* that is the carbon dioxide driving the temperature or vice versa has apparently been solved demonstrating that the temperature change lead by several hundred years the gas changes. A similar behavior is observed for other greenhouse gases like methane. The idea is that the warming during the deglaciation causes the release of the gases from the ocean, soils, and biosphere. On the other hand to explain the lower concentration during the cooling phase one possibility is the enhanced efficiency of the biological pump. During the build up of the ice caps the sea level decreases exposing large portion of the continental platform which releases fresh nutrients to the water with consequence of increasing the transfer of CO_2 from the atmosphere to the ocean. A simple toy model can be developed to reproduce the behavior of carbon dioxide and temperature. We start by saying that the temperature of the planet results from the equilibrium between the solar energy $\bar{S}$, greenhouse effect $\bar{G}$, and the emitted radiation sigma

Fig. 7.20 Result of the toy model for the coupling temperature carbon dioxide

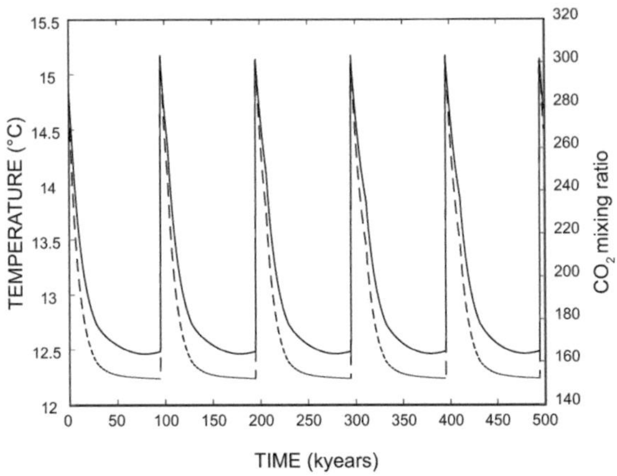

σT^4

$$\bar{S} + \bar{G} = \sigma \bar{T}^4 \tag{7.65}$$

And introducing the heat capacity $c \approx 1.7 \times 10^{10}\text{J/m}^2/\text{K}$ we have

$$c\frac{dT}{dt} = S + G - \sigma T^4 \tag{7.66}$$

Notice the absence of overbars because now those quantities depend on time. The incoming radiation is simply modulated as the astronomical frequency

$$S(t) = \bar{S} + \sum_i S_i \left(\frac{2\pi t}{\Gamma_i}\right) \tag{7.67}$$

where S_i is the perturbation in the incoming radiation and Γ_i is the period. If we use $S_i \approx 0.25\text{w/m}^2$ and $\Gamma_i \approx 10^5$ years we get a perturbation of temperature of the order of 0.1 °C. The greenhouse term relates to the CO_2 mixing ratio C

$$G = \bar{G} + A \ln\left(\frac{C(t)}{C_0}\right) \tag{7.68}$$

where $C_0 = 280ppm$. This equation combined with (7.67) and (7.66) gives

$$c\frac{dT}{dt} = \bar{S} + \sum_i S_i \left(\frac{2\pi t}{\Gamma_i}\right) + \bar{G} + A \ln\left(\frac{C(t)}{C_0}\right) \tag{7.69}$$

We need to specify then the feedback between CO_2 and temperature

$$\frac{dC}{dt} = V - (W_0 + W_1 C) + \beta(C_{max} - C)max\left(\frac{dT}{dt} - \epsilon, 0\right) \tag{7.70}$$

where V is the volcanic source (0.028 ppm/year), the second term is a sink where $W_0 = 0.013 ppm/yr)$ and $W_1 = 8.3 \times 10^{-5}\ yr^{-1}$. The final term represents a release of carbon dioxide with significant warming, i.e., when $dT/dt > \epsilon$, and is limited to a value of C_{max}(governed by the amount of oceanic CO_2 readily available). We take $C_{max} = 400\ ppm$ and $\epsilon = 2.5 \times 10^{-5}$ while $\beta = 0.38\,°C^{-1}$ to take into account the glacial interglacial changes. Equations (7.69)–(7.70) can be integrated and the results are shown in Fig. 7.20. The toy model does not explain the reason for the carbon dioxide feedback that must involve the biogeochemical cycle like described by Paillard and Parrenin (2004).

Appendix

Energy Balance Models

We start from the basic and define a coordinate system where θ is the co-latitude with $\mu = \cos\theta$. Then we assume that heat transport across latitude is of a diffusive type so that the flux and its divergence can be written as

$$q_h = -D\nabla T, \qquad \nabla \cdot q_h = -\nabla \cdot (D\nabla T) \tag{7.71}$$

In spherical coordinate taking just the component in the θ direction we have for the heat flux divergence

$$\nabla \cdot q_h = \frac{d}{d\mu}\left(D \cdot (1-\mu^2)\frac{dT}{d\mu}\right) \tag{7.72}$$

where for the gradient we have used

$$(\nabla)_\theta T = \frac{\sqrt{1-\mu^2}}{a}\frac{dT}{d\mu} \tag{7.73}$$

With a the Earth radius which has been incorporated in the definition of D taken as constant along the latitude. We now assume that the infrared radiation emitted at it is just a linear function of T

$$I(\mu) = A + BT(\mu) \tag{7.74}$$

While the solar radiation Q is distributed in latitude according to some function $S(\mu)$ while the co-albedo is simply $a(\mu) = 1 - \alpha(\mu)$. The energy balance across some latitude can be written as

$$-\frac{d}{d\mu}\left(D\cdot(1-\mu^2)\frac{dT}{d\mu}\right)+A+BT(\mu)=QS(\mu)a(\mu) \tag{7.75}$$

It is convenient at this point to introduce the variable $x=\sin(\pi/2-\theta)=\mu$ so that (7.75) becomes

$$I(x)-\frac{d}{dx}D(1-x^2)\frac{dT(x)}{dx}=QS(x)a(x.x_s) \tag{7.76}$$

whereas we will see in a while a is now function also of the latitude (x_s) where the polar ice will form. Considering the linearity of I(x) we could write

$$I(x)-\frac{d}{dx}D(1-x^2)\frac{dI(x)}{dx}=QS(x)a(x.x_s) \tag{7.77}$$

The solution to this equation can be obtained in a very elegant way (due to G. North) using the fact that the Legendre polynomials are an eigenfunction of the diffusion spherical operator

$$\frac{d}{dx}(1-x^2)\frac{dP_n(x)}{dx}=-n(n+1)P_n(x) \tag{7.78}$$

Some restrictions can be used on terms like $P_n(x)$, because if we consider an annual mean and perfectly symmetrical hemispheres, then the solutions should not change when x changes sign. This means that each variable developed in a series of Legendre polynomials should contain only even terms. For example, for the IR

$$I(x)=\sum_{n\ even} I_n P_n(x) \tag{7.79}$$

That once substituted in (7.77) gives

$$\sum_{n\ even} I_n P_n(x)\,[1+n(n+1)D]=QS(x)a(x,x_s) \tag{7.80}$$

We can now use the orthogonality and normalization properties of the polynomials:

$$\int_0^1 P_n(x)P_m(x)dx=\frac{\delta_{mn}}{1+2n} \tag{7.81}$$

Then if we multiply by (7.80) by $P_m(x)$ integrate will be left only with terms with m = n

$$I_n[1+Dn(n+1)]=(1+2n)Q\int_0^1 P_n(x)S(x)a(x,x_s)dx=QH_n(x_s) \tag{7.82}$$

where

$$H_n(x_s) = (1 + 2n) \int_0^1 P_n(x)S(x)a(x, x_s)dx \tag{7.83}$$

Now we specify the co-albedo

$$a(x, xs) = \begin{cases} b_o, & x > x_s \\ a_0 + a_2 P_2(x), & x < x_s \end{cases} \tag{7.84}$$

where $b_0 = 0.38$, $a_0 = 0.697$, $a_2 = -0,0779$ and perform some arithmetic on the two-mode solution (n=2).

$$\begin{aligned} I_0 &= Q \int_0^1 S(x)a(x.x_s) \\ I_2(1 + 6D) &= QH_2(x_s) \end{aligned} \tag{7.85}$$

Through the second equation we can calibrate D. We fix the present ice limit $x_s = 0.95$ corresponding to latitude of 72°. Now we can approximate $S(x) = 1 + S_2 P_2(x)$ with $S_2 = -0.477$ and obtain $H_o(x_s) = 0.698$ which determine $I_0 = 273.3\ \mathrm{w/m^2}$. Because $I(x_s) = I_0 + I_2 P_2(x_s) = A - 10B = 182.4\ \mathrm{w/m^2}$ from which we get $I_2 = -54.8\ \mathrm{w/m^2}$ so we have

$$T_0 = \frac{I_0 - A}{B} = 33.5\,°\mathrm{C}, \quad T_2 = \frac{I_2}{B} = -27\,°\mathrm{C} \tag{7.86}$$

And the temperature profile $T(x) = T_0 + T_2 P_2(x)$ is determined. Now we can test the sensitivity of the ice line x_s to the change of solar constant. Simply consider the $I_s = I_0 + I_2 P_2(x)$ or more in general

$$I_s = \sum_{neven} I_2 P_n(x)$$

And substituting from (7.82) we have

$$I_s = Q \frac{\sum_{neven} H_n(x_s) P_n(x_s)}{[1 + Dn(n + 1)]} \tag{7.87}$$

Now we can apply this beautiful simple model to different levels of carbon dioxide which presumably will influence A and B in (7.74). Ken Caldeira and James Kasting performed such an effort in Nature paper of 1992 and made a fitting to 2000 runs of a radiative-convective model and came up to express A and B as

$$\begin{aligned} A &= -326.4 + 9.161\varphi - 3.164\varphi^2 + 0.5468\varphi^3 \\ B &= 1.953 - 0.04866\varphi + 0.01309\varphi^2 + 0.002577\varphi^3 \end{aligned} \tag{7.88}$$

where φ is the natural log of atmospheric partial pressure of CO_2 divided by a reference level of 300 ppm. Notice however that A and B obtained from (7.88) use (7.74) with the temperature expressed in °K. In Fig. 7.8 we report the calculation of x_s as function of Q for the present CO_2 partial pressure, for 0.12 and 0.02 bar. We can see that 0.12 bar could avoid the freezing of the Earth even with a reduction of about 30% in the solar constant. In what follow we report a MATLAB script for the two-mode North model

```
xs=[0:0.01:1];
af=0.32
ai=0.68;
a0=0.697;
b0=0.38;
a2=-0.079;
S2=-0.477;
q0=334.4;
D=0.67;
H01=0.5*(xs.^3-xs)*(a2+S2*a0)+0.25*a2*S2*(1.8*xs.^5+xs-2*xs.^2)+a0*xs;
H02=b0*(1-xs-0.5*S2*(xs.^3-xs));
H0=H01+H02
H21=0.5*a0*(xs.^3-xs);
H22=0.25*(a0*S2+a2)*(1.8*xs.^5+xs-2*xs.^3);
H23=0.125*a2*S2*(27*xs.^7/7-5.4*xs.^5+3*xs.^3-xs);
H24=0.25*S2*b0*(0.8-1.8*xs.^5-xs+2*xs.^3);
H2=5*(H21+H22+H23+H24);
P2=0.5*(3*xs.^2-1);
werk1=H0+P2.*H2/(1+6*D);
A=-300;
B=1.9;
IS=A+B*263.15;
q=IS./werk1;
qs1=(q/q0);
plot(qs1,xs,'g')
A=-301.92;
B=1.78;
IS=A+B*263.15;
q=IS./werk1;
qs2=q/q0;
hold on
plot(qs2,xs,'r')
A=-261;
B=1.54
IS=A+B*263.15;
q=IS./werk1;
qs3=q/q0;
hold on
plot(qs3,xs,'k');
MS=[qs1(:),qs2(:),qs3(:),H0(:),xs(:)]
```

Astronomical Forcing

The distance r (see Fig. 7.13) of the Earth from the sun at time t makes an angle ν with the longitude of the perihelion. The Earth orbit is described by the law of mechanics

$$r^2\frac{d\nu}{dt} = a^2(1-e^2)^{1/2} \tag{7.89}$$

where e is the eccentricity of the orbit and a is the semi-major axis, while

$$r(t) = \frac{a(1-e^2)}{1+e\cos[\nu(t)]} \tag{7.90}$$

is the instantaneous distance of the Earth from the sun. In this case, the unit for the time is (1 year)/2π. If we define $\omega = \nu + \Pi$ for a fixed value of the longitude of the perihelion Π we have

$$dt = \frac{r^2 d\omega}{a^2(1-e^2)^{1/2}} \tag{7.91}$$

The solar flux incident at the top of the atmosphere is then

$$S(t, \phi, i, e, \Pi) = \frac{a^2}{r^2}\tilde{S}[\omega(t, e, \Pi), \phi, t] \tag{7.92}$$

where ϕ is the latitude, i is the obliquity, that is, the angle between the normal to the ecliptic plane and the rotation axis of the Earth. S_0 is the solar constant while the quantity

$$\left(\frac{S_0}{4}\right)\tilde{S}[\omega(t, e, \Pi)$$

can be considered as the flux relative to a circular orbit with $r = a$ and $\omega = t$. We already know that

$$\frac{1}{2}\int_{-\pi/2}^{\pi/2} \tilde{S}(\omega, \phi, i) cos\phi d\phi = 1 \tag{7.93}$$

In Fig. 7.21, we show schematically the behavior of the longitude of perìhelion, obliquity, and eccentricity. It is interesting to evaluate the annual average of the solar flux that for an eccentric orbit is given by

$$\frac{S_0}{4}\int_0^{\pi\tilde{S}} \frac{a^2}{r^2}dt = \frac{S_0}{4(1-e^2)^{1/2}}\int_0^{2\pi} \tilde{S} d\omega$$

The annual average is then proportional to the quantity $(1-e^2)^{1/2}$ and does not depend either on the obliquity or on the longitude of the perihelion. However, the annual average radiation at a given latitude other than the eccentricity also depends

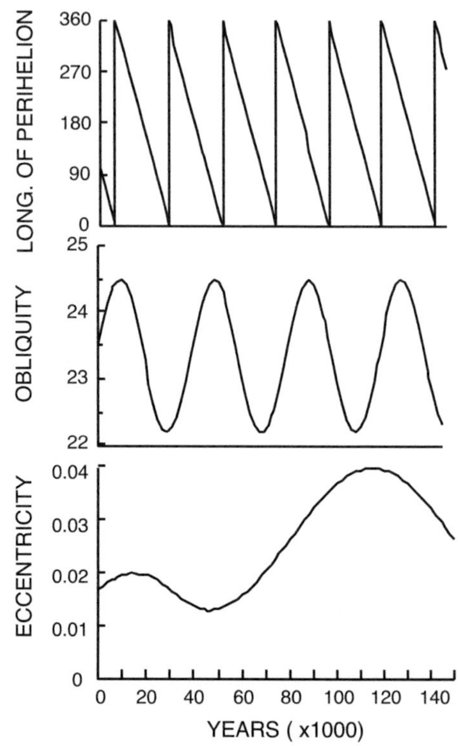

Fig. 7.21 The change with time of longitude of perihelion, obliquity, and eccentricity

weakly on the obliquity (it changes like $\sin(i)$) but again does not depend on the longitude of the perihelion; this, as a matter of fact, only determines the flux distribution during the year at a fixed latitude. If we want to evaluate the absorbed radiation we need to introduce the albedo α so that the radiation absorbed at some latitude is given by

$$\frac{S_0}{4(1-e^2)^{1/2}} \int_0^{2\pi} \tilde{S}\alpha d\omega \tag{7.94}$$

This means that for the absorbed flux to change the albedo must depend on the season (ω). This is the reason why it is so important to have models that can calculate the albedo as a function of the season and not just energy balance models because they must include seasonal effects. The present value of the obliquity is 23° so that the axis is roughly in the middle of its oscillation. On the other hand, the eccentricity is around its minimum value. To have a more precise idea of how the solar flux is influenced by the astronomical parameters we can refer to Fig. 7.22, where the flux difference between two distinct epochs is reported as a function of the latitude and the season. The values shown in the figure are normalized to $S_0/4$. The left part of the figure shows the difference between 10,000 and 30,000 years ago. If we consider Fig. 7.22, we see that the most recent epoch corresponds to the maximum value of the obliquity

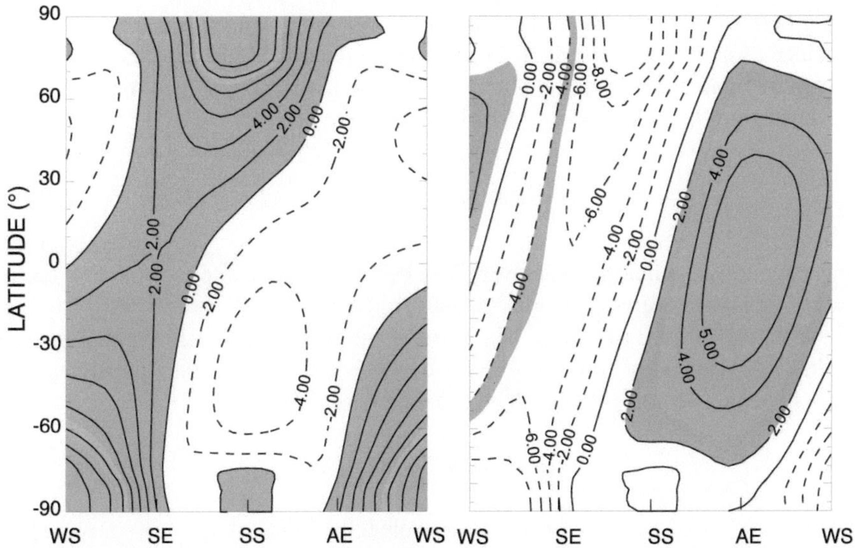

Fig. 7.22 The difference (in %) with respect to the average value of the solar radiation between two different epochs. At left are shown the differences between 10,000 and 30,000 years ago, and on the right between 75,000 and 85,000. Negative values (less radiation at the most recent epoch) are indicated with dashed lines while the positive differences are shaded. WS and SS indicate winter and summer solstice, respectively. SE and AE are summer and autumn equinox

while the most distant epoch corresponds to the minimum. This implies that 10,000 years ago the northern hemisphere in the summer was receiving roughly 12 more solar flux than 30,000 years ago. The two hemispheres are not completely symmetric because the difference has not been chosen to be exactly the period of the equinox precession (23,000 years). The figure on the right shows the same difference between 75,000 and 85,000 years ago. In this case, the obliquity has an opposite effect (the most remote epoch receives more radiation) but now the most important effects are the precession of the equinox and the eccentricity of the orbit. As a matter of fact, the eccentricity increases and so the average annual flux increases, while the 10,000 years difference enhances the effects of the equinox precession. An interesting conclusion of this rather superficial examination is that the seasonal effects are predominant. For thousands of years, some regions of the Earth received a very different quantity of radiation during winter and summer. The simplest mechanism is that the temperature changes that result may change the albedo which amplifies the initial temperature effect (ice–albedo feedback).

Two Stream Approximation

We divide the space in two hemisphere where we assume the radiation fluxes are independent of the direction but are different in the two hemispheres, F^+ in the upper hemisphere and F^- in the lower one. For the downward radiation F^- we consider a fraction f is scattered forward, while a fraction 1–f is scattered backward. These fractions result from the integration of all hemispheres. Then the change in the downward flux when going through the distance dz is

$$\frac{dF^-}{d(-z)} = -a\rho F^- - s\rho(1-f)F^- + s\rho(1-f)F^+ \tag{7.95}$$

This equation says that the downward flux decreases for absorption (the first term) and for the backscattered radiation (second term) and increases for the backscattering of the upward flux (last term). The terms a and s are the absorption and scatter coefficients. The extinction can be defined as k = a + s so that the optical depth is given by $d\tau = -k\rho dz$ with the albedo for single scattering

$$\omega = s/(a+s) = s/k$$

where a,s,k, and ω can be a function of wavelength. Substituting (7.89) becomes

$$\frac{dF^-}{d\tau} + F^- = \omega\left[fF^-(1-f)F^+\right] \tag{7.96}$$

While for the upward flux

$$-\frac{dF^+}{dz} + F^+ = \omega\left[fF^+ - (1-f)F^-\right] \tag{7.97}$$

defining $a_1 = 1 - \omega f, b_1 = \omega(1-f)$ we have eliminating F^-

$$\frac{d^2F^+}{d\tau^2} - \alpha^2 F^+ = 0 \tag{7.98}$$

where $\alpha^2 = a^2 - b^2$. The solution of (7.92) reads

$$F^+ = A\sinh\alpha\tau + B\cosh\alpha\tau \tag{7.99}$$

Now we consider a single layer of optical thickness τ^* and assuming the boundary condition $F^+(\tau^*) = 0$ we have

$$F^+ = \frac{A}{\cosh\alpha\tau^*}(\sinh\alpha\tau\cosh\alpha\tau^* - \cosh\alpha\tau\sinh\alpha\tau^*) = A\sinh\alpha(\tau^* - \tau) \tag{7.100}$$

where the constant A absorbs any other constant. This solution can be used in (7.90) to obtain

$$F^- = \frac{A}{b}[\alpha\cosh(\tau^* - \tau) + a\sinh\alpha(\tau^* - \tau)] \tag{7.101}$$

In order to define the transmission coefficient T and reflection R we define

$$F_{inc} = F^{-}(0), \quad F_{refl} = F^{+}(0), \quad F_{trans} = F^{-}(0), \quad F^{+}(0) = 0.0 \qquad (7.102)$$

so that

$$T = \frac{F_{trans}}{F_{inc}} = \frac{\alpha}{\alpha \cosh \alpha\tau^{*} + a \sinh \alpha\tau}, \qquad R = \frac{F_{refl}}{F_{inc}} = \frac{b \sinh \alpha\tau^{*}}{\alpha \cosh \alpha\tau^{*} + a \sinh \alpha\tau} \qquad (7.103)$$

pH of Acid Rain

As we know the pH of an acid solution is defined as the logarithm of the hydrogen ion concentration. We assume that the hydrogen ions are contributed mainly by sulfuric acid so we consider the dissociation

$$\begin{aligned} H_2SO_4 &\leftrightarrows H^{+} + HSO_4^{-}, \qquad k_a = 10^{-3} \\ HSO_4^{-} &\leftrightarrows H^{+} + SO_4^{2-}, \qquad k_b = 10^{-1.9} \end{aligned} \qquad (7.104)$$

Also, we know the dissociation

$$H_2O \leftrightarrows H^{+} + OH^{-} \qquad (7.105)$$

The charge conservation requires

$$[H^{+}] = [HSO_4^{-}] + 2[SO_4^{2-}] + [OH^{-}] \qquad (7.106)$$

And the total concentration of the sulfur species

$$[H_2SO_4] + [HSO_4^{-}] + [SO_4^{2-}] = 5 \times 10^{-5}\text{mole(S)/liter} \qquad (7.107)$$

where we have used the concentration of the Northeast region of the United States. We have now five Eqs. (7.98–7.101) in five unknowns that we can solve as follows. Eliminate $[HSO_4^{-}]$ between (7.98) and get

$$[H^{+}]^2[SO_4^{2-}] = 10^{1.1}[H_2SO_4] \qquad (7.108)$$

Then obtain $[SO_4^{2-}]$ from (7.102) and the same from (7.98) and (7.99) to have

$$[H^{+}] = \frac{10^{3}[H_2SO_4]}{[H^{+}]} + \frac{10^{1.4}[H_2SO_4]}{[H^{+}]^2} + \frac{10^{-14}}{[H^{+}]} \qquad (7.109)$$

Another equation can be obtained for H_2SO_4

$$[H_2SO_4] = 10^{-4.3}\left[1 + \frac{10^3}{[H^+]} + \frac{10^{1.1}}{[H^+]^2}\right] \tag{7.110}$$

That could be substituted in (7.103) but much more simply we observe that for each three mole of sulfur species there are two moles of [H^+] so we have

$$[H^+] = 2 \times 5 \times 10^{-5} = 10^{-4}, \qquad \text{pH} = 4 \tag{7.111}$$

Jacob in his beautiful little book uses a more straightforward method that has the advantage to take into account other contributions to the acidity besides H_2SO_4. He estimates that, in the USA, each day 1.3×10^9 moles of NO_x, $(NO + NO_2)$ and 2×10^9 moles of SO_2 are released. So, the assumption is that they will form the same amount of $[H^+]$ moles. This means that each mole of sulfur dioxide will release two moles of hydrogen as H_2SO_4, while the nitrogen oxide will release one mole of hydrogen as HNO_3. The total hydrogen released will be

$$2 \times 10^9 + 1.3 \times 10^9 = 3.3 \times 10^9, \text{ moles of } [H^+]$$

These will dilute in the daily volume of rain 2×10^{13} liters/day to reach a concentration of

$$\frac{3.3 \times 10^9}{2 \times 10^{13}} = 1.7 \times 10^{-4} \rightarrow \text{pH} = 3.8$$

A little bit more acid than in the previous calculation.

The pH of the Ocean

Carbon dioxide dissolves in the ocean forming the weak carbonic acid $(CO_2 \cdot H_2O)$ which dissociates in bicarbonate (HCO_3^-) and carbonate (CO_3^{2-}). The definition of equilibrium constant gives

$$\begin{aligned} K_H &= \frac{[CO_2 \cdot H_2O]}{P_{CO_2}} = 3 \times 10^{-2} M/atm \\ K_1 &= \frac{[HCO_3^-][H^+]}{[CO_2 \cdot H_2O]} = 9 \times 10^{-7} M \\ K_2 &= \frac{[CO_3^{2-}][H^+]}{[HCO_3^-]} = 7 \times 10^{-10} M \end{aligned} \tag{7.112}$$

In these equations, P_{CO_2} is the atmospheric pressure of carbon dioxide that is 4×10^{-4} atmosphere and M stands for mole fraction. From these equations we get

$$K_1 \cdot K_2 \cdot K_H = \frac{[CO_3^{2-}][H^+]^2}{P_{CO_2}} \tag{7.113}$$

At this point, we can calculate the pH of the ocean if we assume that the surface water is saturated with respect to C_aCO_3 (calcium carbonate). This compound is one of the products of erosion and we need to add another reaction

$$C_aCO_3 \leftrightarrows Ca^{2+} + CO_3^{2-}, \quad K_S = [Ca^{2+}][CO_3^{2-}] = 9 \times 10^{-7} M^2$$

The concentration of $[Ca^{2+}]$ is 0.01M so that

$$[H^+] = \left(\frac{K_H k_1 K_2}{k_S} [CO_3^{2-}] P_{CO_2}] \right)^{1/2} \tag{7.114}$$

When we substitute the numerical values we get pH≈8.53. It is quite interesting to compare this value of pH with the one of atmospheric water. In this case, we have

$$K_1 \cdot K_H = \frac{[HCO_3^-][H^+]}{P_{CO_2}} \tag{7.115}$$

And now at the equilibrium $[HCO_3^-] \approx [H^+]$ so we have $[H^+]^2 = K_1 \cdot K_H P_{CO_2}$ that is about pH $= 5.5$ so that atmospheric water (what you drink) is more acid than the water in the ocean.

A Pulse of Carbon Dioxide

$$DIC = [H_2CO_3] + [HCO_3^-] + [CO_3^{2-}] \tag{7.116}$$

The concentration of the different forms of carbon can be obtained from the condition that the sea is electrically neutral. Considering the variety of the ions present in the seawater, this is not a simple condition. In particular, ions from strong acids and bases give an excess positive charge that is called alkalinity and is indicated with Alk. Actually, it corresponds to the total charge carried by the carbon compounds

$$Alk = [HCO_3^-] + 2[CO_3^{2-}] \tag{7.117}$$

Consequently we have

$$Alk - DIC = [CO_3^{2-}], \qquad [HCO_3^-] = 2DIC - Alk \tag{7.118}$$

At this point we introduce a solubility constant α such that

$$P_{CO_2} = [CO_2]/\alpha = [H_2CO_3]/\alpha \tag{7.119}$$

And because

$$[H_2CO_3] = \frac{K_2[HCO_3^-]^2}{K_1[CO_3^{2-}]} \tag{7.120}$$

We get

$$P_{CO_2} = \frac{K_2[HCO_3^-]^2}{K_1\alpha[CO_3^{2-}]} = K_{CO_2}\frac{[HCO_3^-]^2}{[CO_3^{2-}]} \tag{7.121}$$

where $K_{CO_2} = K_2/(K_1\alpha)$. Then it is possible to express

$$[HCO_3^-]$$

as a function of Alk and DIC

$$[HCO_3^-] = \frac{\{DIC - [DIC^2 - Alk(2DIC - Alk)(1 - 4K_{CO_2})]^{1/2}\}}{1 - 4K_{CO_2}} \tag{7.122}$$

At this point, we should be able to calculate the partial pressure of carbon dioxide in the atmosphere. We assume that the equilibrium between the dissolved carbon dioxide and the atmosphere is established with a characteristic time given by τ_{dis} so that

$$\frac{dP_{CO_2}}{dt} = \frac{P_{CO_2} - (P_{CO_2})_s}{\tau_{dis}} \tag{7.123}$$

where $(P_{CO_2})_s$ indicates the CO_2 pressure in surface water. Most of the photosynthetic activities happen in the surface water

$$\begin{aligned} 106CO_2 + 16NO_3^- + H_2PO_4^{2-} + 122H_2O \\ \rightarrow (CH_2O)_{106}(NH_3)_{16}(H_3PO_4) + 138O_2 \end{aligned} \tag{7.124}$$

It is quite obvious that in the surface waters, there is a photosynthetic surplus with respect to respiration because a fraction of the material produced sinks to deep waters so that a flux of carbon dioxide is established between the surface and deep ocean. An opposite flux of nutrients is established between the bottom of the ocean and the surface. Surface waters will be impoverished in DIC and enriched in Alk with the opposite happening for deep waters. Nitrates are fixed during photosynthesis and released during respiration. The result of this process (also denoted biological pump) is a depletion of the surface water in carbon dioxide and so of the atmosphere: the higher the biological activity in the ocean, the more severe is the depletion of the surface water. Based on this, we can establish that the change in surface DIC (that we will call DIC_s) is determined by the sedimentation of organic and inorganic carbon and by the exchange between the surface and deep waters due to the circulation. We have then

$$
\begin{aligned}
V_s\frac{dDIC_s}{dt} &= \frac{P_{CO_2}-(P_{CO_2})_s}{\tau_{dis}}C_m-(1+C_{carb})P+(DIC_d-DIC_s)F_w\\
V_d\frac{dDIC_s}{dt} &= (1+C_{carb})P-(DIC_d-DIC_s)F_w
\end{aligned}
\tag{7.125}
$$

where with V_s we have indicated the volume of the upper ocean and with C_m conversion factor between the units of pressure and mass. C_{carb} is the ratio between organic and inorganic carbon, and P is the sedimentation rate of the inorganic carbon. The units for DIC used here are $mole/m^3$ so that the units for P are mole/year. In analogy, DIC_d is the carbon content in deep waters, and F_w is the water flux for the exchange between the surface and deep waters. Exactly in the same manner, we can write two equations for Alk_s and Alk_d considering that, besides the exchange between surface and deep waters, for each mole of carbonate that may end as sediment, there is a change of two equivalent moles of alkalinity due to the term Ca^{2+} which has two positive charges. Each carbon mole is associated, according to Eq. (7.118), with 16/106 $\approx$ 0.15 mol of nitrogen so that we have

$$
\begin{aligned}
\frac{dAlk_s}{dt} &= (Alk_d-Alk_s)F_w-(2C_{carb}-0.15)P\\
\frac{dAlk_d}{dt} &= -(Alk_d-Alk_s)F_w+(2C_{carb}-0.15)P
\end{aligned}
\tag{7.126}
$$

In the MATLAB script that follows we calculate the CO_2 pressure evolution with an initial condition equivalent to 5 PAL (preexisting atmospheric level) as shown in Fig. 7.10 while the other data are as follows: $DIC_s = 2.01 mole/m^3$, $DIC_d = 2.23 mole/m^3$, $Alk_s = 2.2 mole/m^3$, $Alk_d = 2.26 mole/m^3$, $\tau_s = 8.64 year$, $C_m = 4.95\times 10^{16} mole$, $V_s = 1.2\times 10^{17} m^3$, $V_d = 1,23\times 10^{18} m^3$, $P = 1.75\times 10^{14}$ $mole/year$, $C_{carb} = 0.25$, $F_w = 10^{15} m^3/yesr = 31.64 Sverdrup$. The Sverdrup is a unit used by oceanographers and corresponds to a flux of $10^6 m^3/s$.

```
[t,x]=ode45('oceancarb',[0 500],[2.26 2.2 5. 2.247 2.03]);
kcarb=6.35e-4;
hco=(x(:,5)-sqrt(x(:,5).^2-x(:,2).*(2*x(:,5)-x(:,2))*(1-4*kcarb)))/(1-4*kcarb);
co3m=(x(:,2)-hco)/2.;
kh=3.e-2;
k1=9.e-7;
k2=7.e-10;
wkph=sqrt(k1*k2*kh*x(:,3)./co3m);
pH=-log10(wkph);
figure (1)
loglog(t,x(:,3),'k',t,x(:,4),'r',t,x(:,5),'g'),axis([0 500 0 5])
figure(2)
semilogx(t,pH),axis([0 500 0 10])
function xdot=oceancarb(t,x)
%alkd=x(1)
%alks=x(2)
%pco2=x(3)
%sigd=x(4)
%sigs=x(5)
xdot=zeros(5,1);
watem=288;
```

```
vols=1.2e17;
vold=1.23e18;
corat=0.25;
kcarb=5.75e-4+6.e-6*(watem-278);
kco2=0.035+0.0019*(watem-278);
matmco2=5.02e16;
prod=1.75e14;
proda=(2*corat-0.15)*prod;
prods=(1+corat)*prod;
dstime=8.64;
wflux=0.001e18;
%if t<250
%antr1=3e12*exp(t/26.9);
%antr=antr1/(44*matmco2);
%else
antr=0
%end
dalk=(x(1)-x(2))*wflux;
dsig=(x(4)-x(5))*wflux;
hco3=(x(5)-sqrt(x(5)^2-x(2)*(2*x(5)-x(2))*(1-4*kcarb)))/(1-4*kcarb);
co3=(x(2)-hco3)/2.;
pco2s=kco2*hco3*hco3/co3;
xdot(1)=(proda-dalk)/vold;
xdot(2)=(dalk-proda)/vols;
xdot(3)=(pco2s-x(3))/dstime+antr;
xdot(4)=(prods-dsig)/vold;
xdot(5)=((-(pco2s-x(3))*matmco2/dstime+dsig)-prods)/vols;
```

References

Alvarez, L.W. et al.: Extraterrestrial cause for the Cretaceous–Tertiary extinction. Science **208** (1980)

Benzi, R., et al.: A theory of stochastic resonance in climatic change. SIAM J. Appl. Math. **43**, 565 (1983)

Birks, J.W. et al.: Frequent ozone depletion resulting from impacts of asteroids and comets. In: Bobrowsky, P., Rickman, H. (eds.), Comet/Asteroid Impacts and Human Society. Springer (2005)

Broecker, W.J.: Glacial World according to Wally. Eldigio Press (2002)

Broyles, A.A.: Smoke generation in a nuclear war. Am. J. Phys. **53**, 323 (1985)

Budyko, M.I.: The effect of solar radiation variations on the climate of the Earth. Tellus **21**, 611 (1969)

Caldeira, K., Kasting, J.F.: Susceptibility of the early Earth to irreversible glaciation caused by carbon dioxide clouds. Nature **359**, 226 (1992)

Caldeira, K., Kasting, J.F.: The life span of the biosphere revisited. Nature **360**, 721 (1992)

Catling, D.C., Zahnle, K.J.: The Archean atmosphere. Sci. Adv. **6**, eaax1420 (2020)

Charnay, B., et al.: Is the faint young Sun problem for Earth solved? Space Sci. Rev. **216**, 1 (2020)

Crowley, J.T.: G.R. North, Paleoclimatology, Oxford University Press (1996)

Crutzen P.J., Birks, J.W.: The atmosphere after a nuclear war: twilight at noon. Ambio. **11.114** (1982)

Doney, S.C., et al.: Impact of anthropogenic atmospheric nitrogen and sulfur deposition on ocean acidification and the inorganic carbon system. Proc. Natl. Acad. Sci. USA **104**, 14580 (2004)

Donnadieu, Y., Goddéris, Y. et al.: A 'snowball Earth' climate triggered by continental break-up through changes in runoff. Nature **428**, 303 (2004)

Goosse, H.: Climate System Dynamics and Modelling. Oxford University Press (2015)

Gutierrez-Zalapa, R.: Modeling the impact of asteroids over Mexican territory. Revista Mexicana de Fısica **70**, 0107 (2024)

Hansen, J., et al.: Climate change and trace gases. Phil. Trans. R. Soc. A **365**, 1925 (2007)

Henehan, M.J., et al.: Rapid ocean acidification and protracted Earth system recovery followed the end-Cretaceous Chicxulub impact. Proc. Natl. Acad. Sci. USA **116**, 22500 (2019)

Hogg, M.: Glacial cycles and carbon dioxide: a conceptual model. Geophys. Res. Lett. **35**, L01701 (2008)

Jacob, D.J.: An Introduction to Atmospheric Chemistry. Princeton University Press

Kasting, J.F.: Bolide impact and the oxidation state of carbon in the Earth's early atmosphere. Orig. Life Evol. Biosph. **20**, 199 (1990)

Lewis, J.S., et al.: Chemical consequences of major impact events on Earth. Geological Society of America Special Paper, vol. 190 (1982)

Lin, J., Qian, T.: Earth's climate history from 4.5 billion years to one minute. Atmos.-Ocean **60**, 188–232 (2022)

Marshall, J.E.A. et al.: UV-B radiation was the Devonian-Carboniferous boundary terrestrial extinction kill mechanism. Sci. Adv. **6**, eaba0768 (2020)

Morgan, J.V., et al.: The Chixculub impact and its environmental consequences. Nat. Rev.: Nat. Environ. **3**, 338 (2022)

Newmann, M.J., Rood, R.T.: Implications of solar evolution for the earth's early atmosphere. Science **198**, 1035 (1977)

Paillard, D.: The timing of Pleistocene glaciations from a simple multiple-state climate model. Nature **391**, 378 (1998)

Paillard, D.: Glacial cycles: towards a new paradigm. Rev. Geophys. **39**, 325 (2001)

Paillard, D., Parrenin, F.: The Antarctic ice sheet and the triggering of deglaciations. Earth Planet. Sci. Lett. **227**, 263 (2004)

Penman, D.E.: Silicate weathering as a feedback and forcing in Earth's climate and carbon cycle. Earth Sci. Rev. **209**, 103298 (2020)

Pierazzo, P., et al.: Ozone perturbation from medium-size asteroid impacts in the ocean. Earth Planet. Sci. Lett. **299**, 263 (2010)

Pollack, J.B., et al.: Environmental effects of an impact-generated dust cloud: implications for the cretaceous-tertiary extinctions. Science **219**, 287 (1983)

Prinn, R.G., Fegley, B.: Bolide impacts, acid rain, and biospheric traumas at the Cretaceous-Tertiary boundary. Earth Planet. Sci. Lett. **83**, 1 (1987)

Raymo, M.E., Huybers, P.: 2008, Unlocking the mysteries of the Ice Ages. Nature **451**, 284 (2000)

Rial, J.A.: Pacemaking the ice ages by frequency modulation of earth orbital eccentricity. Science **285**, 564 (1999)

Rosing, M.T., et al.: No climate paradox under the faint early Sun. Nature **464**, 744 (2010)

Sagan, C., Mullen, G.: Earth and Mars: evolution of atmospheres and surface temperatures. Science **177**, 52–56 (1972)

Saltzmann, B.: Dynamical Paleoclimatology: Generalized Theory of Global Climate Change. Academic Press (2001)

Santa Catharina, A. et al.: Timing and causes of forest fires at the K-Pg boundary. Sci. Rep. **12**, 13005 (2022)

Sellers, W.D.: A global climatic model based on the energy balance of the Earth-atmosphere system. J. Appl. Meteorol. **8**, 392 (1969)

Toon, O.B., et al.: Evolution of an impact-generated dust cloud and its effects on the atmosphere. Geol. Soc. Am. Spec. Pap. **190**, 187 (1982)

Turco, R.P., et al.: Nuclear winter: global consequences of multiple nuclear explosions. Science **222**, 1283 (1983)

Tyrrell, T., et al.: Severity of ocean acidification following the end-Cretaceous asteroid impact. Proc. Natl. Acad. Sci. USA **112**, 6556 (2015)

Valley, J.W., Peck, W.H., King, E.M.: A Cool Early Earth Geol. **30**, 351 (2002)

Walker, J.C.G., et al.: A negative feedback mechanism for the long-term stabilization of Earth surface temperature. J. Geophys. Res. **86**, 9776 (1981)

Watson, E.B., Harrison, T.M.: Zircon thermometer reveals minimum melting conditions on earliest earth. Science **308**, 841 (2005)

Watts, R.G., Hayder, E.: The origin of the 100-Kiloyear ice sheet cycle in the pleistocene. J. Geophys. Res. **88**, 5163 (1983)

Willeit1, M. et al.: Mid-Pleistocene transition in glacial cycles explained by declining CO_2 and regolith removal. Sci. Adv. **5**, eaav7337 (2019)

Zahnle, K.J.: Atmospheric chemistry by large impacts. Geol. Soc. Am. Spec. Paper **247** (1990)

Zalasiewicz, J., Williams, M.: A geologic history of climate change, in Climate Change: Observed impact of planet Earth, Elsevier (2009)

Zalasiewicz, J.A., Williams, M.: Climate change through Earth history, 3rd edn. Climate Change, Elsevier (2021)

Chapter 8
The Present Status of Atmosphere and Oceans

We are confident we gave some idea how the fluid portions of the planet have evolved in more than 4.5 billion years. It is time to ask what are today the conditions of the atmosphere and the oceans while we reserve the last chapter to make some hypotheses on what to expect in the future. The first thing to notice is that for the first time in the history of our planet both the atmosphere and the oceans feel the presence of human activity. We all know the problem related to the emission of greenhouse gases (mainly carbon dioxide) for the atmosphere and the acidification of oceans and rain. We have also mentioned the slow increase in temperature of the surface ocean, the plastic pollution, and so on.

8.1 The Present Atmosphere

The mass of the present atmosphere amounts to 5.15×10^{18} kg and as we have seen is made up mainly of nitrogen (N_2) and oxygen (O_2) which together account for the 99% of the mole fraction (the ratio of the number of molecules of a given gas to the total number of molecules of air). Table 8.1 gives the details of the present atmospheric composition. Except for the first two components, the rare gases and hydrogen, all the other concentrations are changing due to human activity. This is one of the manifestations of the new geologic epoch which has been baptized *Anthropocene*. As we have mentioned gases like carbon dioxide, methane, and nitrous oxide are so-called greenhouse gases because they maintain the surface temperature of the Earth to reasonable values for the life to exist. Some of these gases like nitrous oxide are responsible for the production of oxide of nitrogen (NO_x) which have effects on the atmospheric ozone.

We do not mention here important gases like water vapor (H_2O) whose mass mixing ratio may reach values of the order 10^{-2} and that can be regarded as the most important greenhouse gas and together with methane a very important element in the

G. Visconti, *Origin and Evolution of the Atmosphere and Ocean*,
https://doi.org/10.1007/978-3-032-02653-8_8

Table 8.1 Most abundant atmospheric gases

Gas	Mole fraction	Mass mix ratio	Mass (kg)
Nitrogen (N_2)	0.78	0.75	3.86×10^{18}
Oxygen (O_2)	0.21	0.23	1.18×10^{18}
Argon (Ar)	0.0093	0.012	6.6×10^{16}
Carbon dioxide (CO_2)	4.30×10^{-4}	6.5×10^{-4}	3.26×10^{15}
Neon (Ne)	1.8×10^{-5}	1.25×10^{-5}	6.43×10^{13}
Ozone (O_3)	$0.01–10 \times 10^{-6}$	$0.016–16.5 \times 10^{-6}$	$0.08\text{-}8.5 \times 10^{13}$
Helium (He)	5.2×10^{-6}	0.7×10^{-6}	3.6×10^{12}
Methane (CH_4)	1.9×10^{-6}	1.0×10^{-6}	5.15×10^{12}
Kripton (Kr)	1.11×10^{-6}	2.36×10^{-6}	1.21×10^{13}
Hydrogen (H_2)	0.53×10^{-6}	3.6×10^{-8}	1.85×10^{11}
Nitrous oxide (N_2O)	3.40×10^{-7}	5.15×10^{-7}	2.65×10^{12}

chemistry of hydroxil (HO_x) and consequently ozone. Other components absent in the table are the class of chlorofluorocarbon (CFC) and hydrochlorofluorocarbons (HCFC) with much smaller concentration (down to part per trillion (ppt)) but which were responsible for the alarm on the ozone depletion and were banned by the Montreal protocol and more recent treatises. From this table as an example we can deduce that 1 ppm of CO_2 is equivalent to about 7.6×10^{12} kg considering that in a period of about 150 years the mixing ratio goes from 280 to 430 ppm the total amount emitted in the atmosphere is about 1.14×10^15 kg that is 1.140 Petagram per year. Actually this underestimates the real emission rate because part of the gas is absorbed by the ocean.

8.2 The Anthropogenic Gases

We will include in this discussion carbon dioxide (CO_2), methane (CH_4), and nitrous oxide (N_2O). We will then discuss ozone (O_3).

Carbon dioxide, CO_2

At the present time (2024), the annual emission rate for carbon dioxide amounts to about 41.6 Gt of which 37.4 Gt are from combustion and the rest from other practices. Not all of the emitted CO_2 goes into the atmosphere because in part is absorbed by the ocean in a ratio which is known as *airborne fraction*. This can be estimated from both the table and some additional information. From the table, we can estimate that we need an average of about $3260/430 \approx 7.6$ Gt of CO_2 per 1ppm. From 2024 to 2024, the total amount of emitted CO_2 can be estimated to be about 357 Gt that would imply an increase of $357/7.6 \approx 47$ ppm while the observed one for the same period is 24 ppm. The airborne fraction is then so $24/47 \approx 0.51$. We can actually

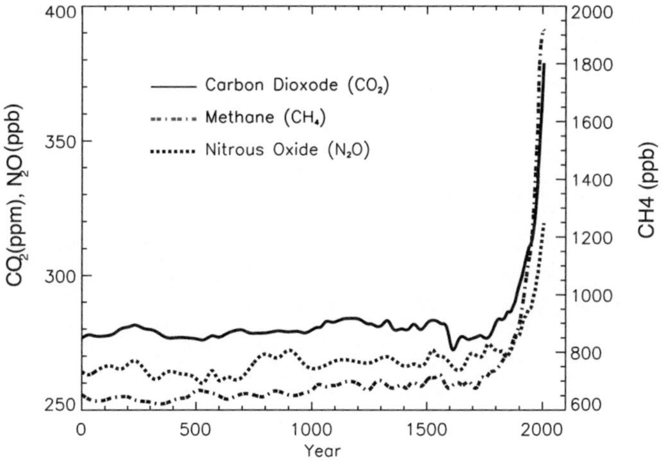

Fig. 8.1 The history of carbon dioxide, nitrous oxide, and methane

calculate such number and this is done in the appendix. Data on the atmospheric mixing ratio of carbon dioxide are taken since the year 1951 and show an increase depicted in Fig. 8.1. These data show a constant increase from about 280 pp to the present 440 pm. The main driver of the increase is the fossil fuel usage.

Methane, CH_4

The main sources of methane are listed in Table 8.2. The items, oil, gases, and coal, are related to the extraction processes and transport. However the sources are mostly related to anthropogenic activities. The main sink for methane is reduction to CO_2 through interaction with hydroxyl, OH (see Appendix).

Methane concentration follows a large increase after the preindustrial times in a very similar fashion to the carbon dioxide. Since then the mixing ratio has increased by a factor of 2.5 while its natural sources and sinks changed only 3%. On the other end, the same mixing ratio has changed from about 1.6 to 1.85 in the last 40 years as shown in Fig. 8.1. Because one of the major sources is related to livestock the increase is referred also as *hamburger connection*: trees which absorb CO_2 are cleared and substituted with pastures for livestock. Also in the last few decades the extractive practice of fracking has to be mentioned.

Nitrous Oxide (N_2O) As we have seen for methane also nitrous oxide is strongly related to agricultural activity. Table 8.3 taken from Tian et al. is a list of sources and sinks. As results from the data the anthropogenic sources account for a little bit less than 40% of the total emission. Nitrous oxide can result from various soil management practices like use of fertilizers, the management of manure, or burning of agricultural residues. The natural occurring emission is mostly from tropical soils because they have higher nutrient availability and moisture levels. This facilitates microbial nitrification and denitrification. Much of the nitrous oxide in the oceans arises from microbial activity in and around sinking particles. These particles provide

Table 8.2 Sources of methane (in Tg/year)

Natural	
Wetlands	160
Open fires	17
Termites and other	5
Subtotal	182
Anthropogenic	
Livestock	110
Oil and gas	70
Waste	60
Coal	50
Rice	40
Other	40
Subtotal	370
Grand total	552±60

the anaerobic conditions necessary for denitrification, a process which creates nitrous oxide as a by-product.

For what concerns the fossil fuels, nitrous oxide is a product of combustion even though depends on the type of fuel and combustion technology, maintenance, and operating practices. In the chemical industry N_2O is emitted during the production of

Table 8.3 Sources and sinks of nitrous oxide (in Tg/year)

Natural	
Soil	8.9
Ocean	5.30
Lightning	1
Coastal waters	0.35
Surface sink	−0.1
Subtotal	15.4
Anthropogenic	
Soil emission	3.6
Manure	2.2
Aquaculture	0.1
Fossil fuels	2.2
Waste	0.5
Biomass burning	0.9
Subtotal	9.5
Grand total	24.9

chemical like nitric acid, nitrous oxide is also generated from treatment of domestic wastewater during nitrification and denitrification of the nitrogen present, usually in the form of urea, ammonia, and proteins. Before the rise of modern agricultural practices and industrial activities the N_2O was in equilibrium being destroyed in the stratosphere with reaction $N_2O + O(^1D) \rightarrow 2NO$. It has been estimated that the total chemical sink is around 21 Tg/year of N_2O while we see from Table 8.3 that the total production amounts to 24.9 Tg/year so that the surplus is $\approx$3.9 Tg/year which produces a constant increase in the nitrous oxide mixing ratio from 270 part per billion (ppb) before the industrial revolution to present values of 340 ppb. The residence time of nitrous oxide in the atmosphere is around 120 years much longer than methane which is roughly 8 years. In conclusion, some of these gases are of natural origin while the human activities have unbalanced their natural cycle determining a constant rise in their mixing ratio.

8.3 Radiative Forcing and Climate

One of the effects of the anthropogenic gases is the additional radiative forcing of climate. In very simple terms if the incoming solar flux is F_s and the Earth has an albedo α infrared radiation is in equilibrium must be $F_s(1-\alpha)/4$. When we add a mass Δm of an absorbing gas the outgoing terrestrial radiation flux will decrease by an amount ΔF so that the temperature of the planet should change accordingly to maintain the equilibrium. Methods have been developed to evaluate the radiative forcing. This is done using very detailed line-by-line calculations. Building on the pioneering work of Gunnar Myhre these calculations are parameterized in terms of the concentration of the relevant gas. We will report the final formulas developed for carbon dioxide, methane, and nitrous oxide. Following the notation of the most recent work we will indicate with upper case C the carbon dioxide mixing ratio (in ppm), with upper N the mixing ratio of nitrous oxide N_2O (in ppb) and with upper case M the mixing ratio (in ppb) of methane CH_4. The expression for carbon dioxide is

$$\Delta F_{CO_2} = [a_1(C-C_0)^2 + b_1(C-C_0) + c_1\bar{N} + 5.36] \ln\left(\frac{C}{C_0}\right) \tag{8.1}$$

where $a_1 = -2.41 \times 10^{-7}$watt m^{-2} ppm^{-2}, $b_1 = 7.2 \times 10^{-7}$watt m^{-2} ppm^{-1}, $c_1 = -2.1 \times 10^{-7}$watt m^{-2} ppb^{-1}. Also C_0 is the initial concentration. Notice that the formula do imply a possible decrease in the mixing ratio but also the dependence on the nitrous oxide concentration, $\bar{N}$. This is due to the partial overlapping of the $17cm^{-1}$ of N_2O with the $15cm^{-1}CO_2$ band. For N_2O and CH_4 we have similar formula

$$\Delta F_{N_2O} = [a_2\bar{C} + b_2\bar{N}c_2\bar{M} + 0.117](\sqrt{N} - \sqrt{N_0}) \tag{8.2}$$

$$\Delta F_{CH_4} = [a_3\bar{M} + b_3\bar{N} + 0.043](\sqrt{M} - \sqrt{M_0}) \tag{8.3}$$

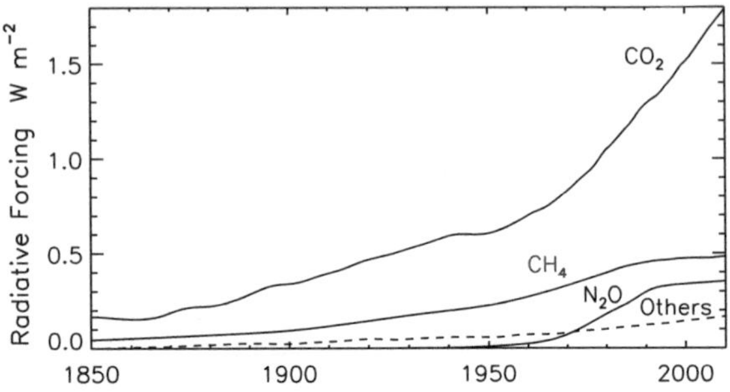

Fig. 8.2 The history of radiative forcing

Again N_0 and M_0 are the initial concentration and $a_2 = -8.0 \times 10^{-6}$ *watt* m^{-2} ppm^{-1}, $b_2 = 74.2 \times 10^{-6}$ *watt* m^{-2} ppb^{-1}, $c_2 = 4.9 \times 10^{-6}$ *watt* m^{-2} ppb^{-1}, $a_3 = -1.3 \times 10^{-6}$ *watt* m^{-2} ppb^{-1}, $b_3 = 8.2 \times 10^{-6}$ *watt* m^{-2} ppb^{-1}. These simple fitting formulas could be used with data on concentration to produce the historical forcing as shown in Fig. 8.2 that demonstrates how the radiative forcing depends on the quantity of greenhouse gas which resides in the atmosphere.

8.4 The Atmospheric Ozone

The rise of ozone in the Earth's atmosphere is a consequence of the rise of oxygen. This can be easily seen from the simple Chapman model which considers a pure oxygen atmosphere and then the reactions

$$\begin{aligned} O_2 + h\nu &\rightarrow O + O, \qquad J_2 \\ O_2 + O + M &\rightarrow O_3 + M, \qquad k_1 \\ O_3 + h\nu &\rightarrow O + O_2, \qquad J_3 \\ O_3 + O &\rightarrow 2O_2, \qquad k_2 \end{aligned} \tag{8.4}$$

We now simply consider production and loss of odd oxygen

$$J_2[O_2] = k_2[O_3][O] \tag{8.5}$$

While the ratio between $[O_3]$ and [O] is given by the second and third reactions

$$k_1[O_2][O][M] = J_3[O_3]$$

Combined with the previous relation, we get

$$[O_3] = [O_2]\left(\frac{J_2 k_1 [M]}{J_3 k_2}\right)^{1/2} \tag{8.6}$$

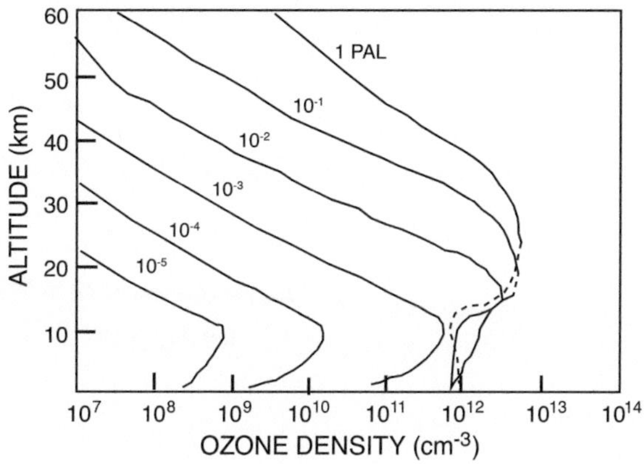

Fig. 8.3 The ozone profiles for different PALs (present atmospheric levels) of oxygen

This simple theory was formulated in the 30s of the last century and does not give the right numbers because overestimate the number density of ozone in the present situation (1PAL). However give some indication on the behavior of ozone especially in the notation of odd oxygen ($0 + O_3$) which is exemplified by the relation (8.6) so that ozone density is proportional to the oxygen density and to its photodissociation while the distribution with altitude depends on the ozone photodissociation. The results taken from different models are shown in Fig. 8.3 for different oxygen levels. The Chapman model overestimates the quantity of ozone because it neglects the existence of the catalytic cycles that follow the scheme

$$\begin{aligned} X + O_3 &\rightarrow XO + O_2 \\ XO + O &\rightarrow X + O_2 \\ \text{net,} \quad O_3 + O &\rightarrow 2O_2 \end{aligned} \tag{8.7}$$

The cycling between X and XO continues until one of them is concerted into a reservoir species. In the modern Earth atmosphere, the most efficient catalytic cycle is based on the hydroxyl (OH and HO_2) and nitrogen oxides (NO and NO_2). Production of OH follows the scheme

$$H_2O + O(^1D) \rightarrow OH + OH \tag{8.8}$$

where $O(^1D)$ is produced by the photodissociation of ozone. The OH radical is a major sink for multiple atmospheric gases (e.g., CH_4, CO) and is often called the "detergent of the atmosphere" for this reason. In a similar manner NO_x are produced by

$$N_2O + O(^1D) \rightarrow NO + NO \tag{8.9}$$

In addition to the catalytic cycle

$$\begin{aligned} OH + O_3 &\rightarrow HO_2 + O_2 \\ HO_2 + O &\rightarrow OH + O_2 \\ \text{net } O_3 + O &\rightarrow 2O_2 \end{aligned} \tag{8.10}$$

Ozone destruction also follows the other path

$$\begin{aligned} OH + O_3 &\rightarrow HO_2 + O_2 \\ HO_2 + O_3 &\rightarrow OH + 2O_2 \\ \text{net } O_3 + O_3 &\rightarrow 3O_2 \end{aligned} \tag{8.11}$$

Or again

$$\begin{aligned} OH + O &\rightarrow H + O_2 \\ H + O_2 + M &\rightarrow HO_2 + M \\ HO_2 + O &\rightarrow OH + O_2 \end{aligned} \tag{8.12}$$

In a similar way, the NO_x catalytic cycle

$$\begin{aligned} NO + O_3 &\rightarrow NO_2 + O_2 \\ NO_2 + O &\rightarrow NO + O_2 \end{aligned} \tag{8.13}$$

However, in certain conditions, in the troposphere NO_x may contribute to ozone formation in the process of oxidation of carbon monoxide (CO)

$$\begin{aligned} CO + OH &\rightarrow CO_2 + H \\ H + O_2 + M &\rightarrow HO_2 + M \end{aligned} \tag{8.14}$$

In presence of NO_x, an alternate reaction is

$$HO_2 + NO \rightarrow OH + NO_2 \tag{8.15}$$

which regenerates OH producing also NO_2 which can photolyze and giving rise to ozone

$$NO_2 + h\nu \xrightarrow{O_2} NO + O_3 \tag{8.16}$$

The net reaction between (8.14) and (8.16) is then

$$CO + 2O_2 \rightarrow CO_2 + O_3 \tag{8.17}$$

The chain reaction initiates with the formation of hydroperoxyl HO_2 which can be destroyed by

$$HO_2 + HO_2 + H_2O_2 + O_2 \tag{8.18}$$

where hydrogen peroxide is rained out or photodissociated or consumes OH

$$
\begin{aligned}
H_2O_2 + h\nu &\rightarrow 2OH \\
H_2O_2 + OH &\rightarrow HO_2 + H_2O
\end{aligned}
\tag{8.19}
$$

The efficiency of ozone production depends on the NO_x presence. The difference between the destroying HO_x catalytic cycles in the stratosphere and the opposite effect in the troposphere is due to the fact that in the latter region the concentration of O atoms is much lower to such an extent to slow down the closing of the cycle. The presence of hydrocarbons in polluted atmosphere could enhance the production of ozone.

8.5 The Present Status of the Ocean

In this is rather impossible task we will follow the IPCC reports (AR5,AR6). We may observe that the most relevant features affected by the climate changes are the salinity of the waters, their pH, the temperature, and the sea level which depends mostly from the extra heat absorbed together and the melting of ice sheets. The status of the ocean will be described mainly as determined by the last century of human-induced global change.

8.5.1 *The Sea Level and the Energy Stored*

We have the ideas considering an imbalance in the radiative flux can influence the temperature and then the sea level. Consider a water column as in Fig. 8.4 of mass δm given by

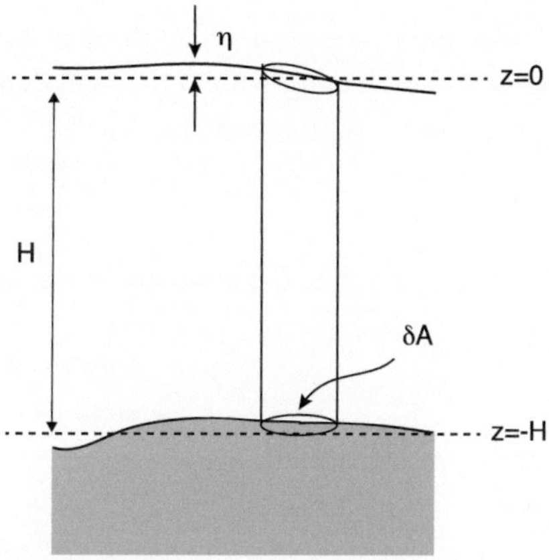

Fig. 8.4 The computation of steric level

$$\delta m = \delta A \int_{-H}^{\eta(t)} \rho(z, t) dz \tag{8.20}$$

where -H is the depth of the base, $\rho(z, t)$ is the density of the water, δA is the area of the mass element, and $\eta(t)$ is the deviation of the sea level. Deriving (8.20) with respect to time we have

$$\frac{\partial(\delta m)}{\partial t} = \delta A \left[\rho(\eta, t) \frac{\partial \eta}{\partial t} + \int_{-H}^{\eta(t)} \frac{\partial \rho}{\partial t} dz \right] \tag{8.21}$$

where now $\rho(\eta, t)$ is the surface density ρ_s so we have

$$\frac{\partial \eta}{\partial t} = -\frac{1}{\rho_s} \int_{-H}^{\eta(t)} \frac{\partial \rho}{\partial t} dz + \frac{1}{\rho_s} \frac{\partial}{\partial t} \left(\frac{\delta m}{\delta A} \right) \tag{8.22}$$

This gives the change in the ocean level as a function of time which results from a steric contribution due to the changes of density and a contribution due to the changes of mass (called mass component). The latter may be produced by the melting of alpine ice masses or the Antarctic Continent. We can consider that the water column is comprised between the atmospheric pressure p_{atm} and the pressure at the bottom p_b so we can write $p_b = p_{atm} + g(\delta m/\delta A)$. In this case (8.22) becomes

$$\frac{\partial \eta}{\partial t} = -\frac{1}{\rho_s} \int_{-H}^{\eta(t)} \frac{\partial \rho}{\partial t} dz + \frac{1}{g\rho_s} \left(\frac{\partial p_b}{\partial t} - \frac{\partial p_{atm}}{\partial t} \right) \tag{8.23}$$

The changes in density can be reduced to temperature changes (*thermosteric component*) or change in salinity (*halosteric component*) so that referring to the equation of state we write

$$-\frac{\Delta \rho}{\rho} = \alpha_T \Delta T - \beta \Delta S \tag{8.24}$$

At this point, we assume that the density variations are related to the deviation of sea level and indicating with D the total depth (8.24) becomes

$$\frac{\Delta \eta}{D} = \alpha_T \Delta T \tag{8.25}$$

We can now relate the anomaly of the heat flux ΔF to the heat content to have

$$\rho C_p D \Delta T = \Delta F \Delta t \tag{8.26}$$

And (8.25) becomes

$$\frac{\Delta \eta}{\Delta t} \approx \frac{\alpha_T \Delta F}{\rho C_p} \tag{8.27}$$

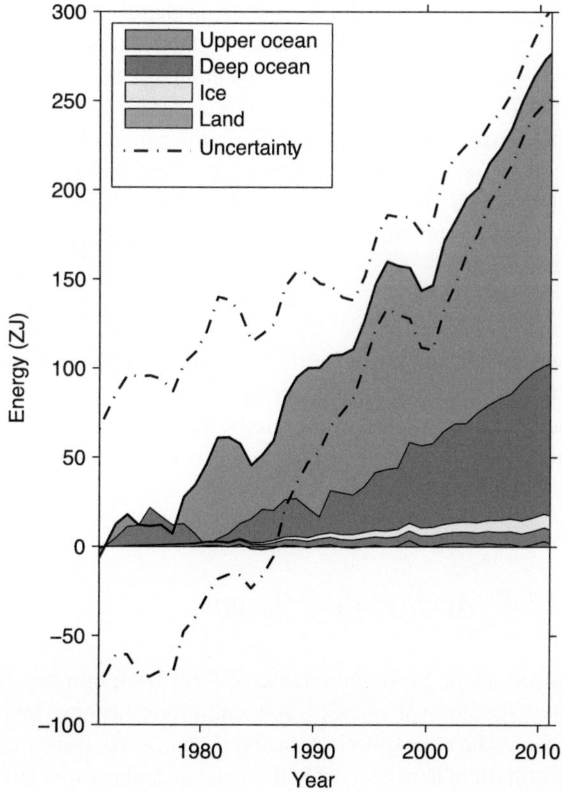

Fig. 8.5 The energy stored by different reservoirs since early 60s

With values $\alpha_T = 1.57 \times 10^{-4} K^{-1}$, $C_p \approx 4 \times 10^3 J K^{-1}\mathrm{kg}^{-1}$, $\rho \approx 10^3 \mathrm{kg\ m}^{-3}$ we find $\alpha_t/\rho C_p \approx 1.26 mm/year$. In the period (1961–2008), the flux change is $0.53\mathrm{wm}^{-2}$ so that for the same period we can get a change of the order of 0.66 mm/year. As for the heat content measurements are available since 1955 and in Fig. 8.5 we show the change in the heat content since 1970. We see that in the first 2000 m of ocean the content rises about 300 zettajoules (1 zettajoule (ZJ) = 10^{21} J).

In the first 700 m it reaches about 200 ZJ. It is interesting to note that the heat absorbed by the ocean amounts to about 90 % of the excess heat due to global warming. Considering that the ocean surface is about 70% of the surface of the Earth that is $5.1 \times 10^{14}\ \mathrm{m}^2$ so that the first 2000 m corresponds to volume of $1.02 \times 10^{18}\mathrm{m}^3$ with a total heat capacity of $\rho C_p \Delta V \approx 4 \times 10^{24}\ J K^{-1}$. Now from 1990 to 2020 we see that the heat content has increased by about 250 ZJ at a rate of 8.3 ZJ/year that means

$$\frac{8.3 \times 10^{21}}{5.1 \times 10^{14} \cdot 3.1 \times 10^{7}} \approx 0.5 \mathrm{w\ m}^{-2}$$

which is the rate we mentioned before. Over the period 1971–2018 (2010–2018), the majority of heat gain is reported for the global ocean with 89% (90 %), with 52%

for both periods in the upper 700 m depth, 28% (30%) for the 700–2000 m depth layer and 9% (8%) below 2000 m dept. In contrast heat gain over land amounts to 6% over these periods, 4% is available for the melting of grounded and floating ice, and 1% is available for atmospheric warming. This Earth energy imbalance (EEI) is the most critical number defining the prospects for continued global warming and climate change and depends on the fact that the absorbed energy is greater than the emitted one. Conceptually, the relationships between EEI, radiative forcing, and surface temperature change can be expressed as

$$\Delta N_{TOA} = \Delta F_{ERF} - |\alpha_{FP}|\Delta T_S \tag{8.28}$$

where ΔN_{TOA} is the Earth's net energy imbalance at TOA (top of the atmosphere) in wm^{-2},ΔF_{ERF} is the net radiative forcing (wm^{-2}), ΔT_S is the global surface temperature anomaly (K) relative to the equilibrium state, and α_{FP} is the net total feedback parameter (wm^{-2} K^{-1}), which represents the combined effect of the various climate feedbacks.

8.5.2 *Acidity of the Oceans*

Since about 1750, the release of CO_2 from human activities has resulted in global average atmospheric CO_2 concentrations that have increased from 278 to 420 ppm in 2024. The atmospheric concentration of CO_2 is now higher than experienced on the Earth for at least the last 800,000 years and is expected to continue to rise because of our dependence on fossil fuels for energy. To date, the oceans have absorbed roughly one-fourth of the total amount of CO_2 emitted by human activities since preindustrial times. This natural process of absorption has significantly reduced the greenhouse gas levels in the atmosphere and minimized some of the impacts of global warming. However, the ocean's uptake of CO_2 is having a significant impact on the chemistry of seawater. The average pH of ocean surface waters has already fallen by about 0.1 units, from about 8.2 to 8.1 since the beginning of the Industrial Revolution. Estimates of projected future atmospheric and oceanic CO_2 concentrations indicate that, by the end of this century, the average surface ocean pH could be 0.2 to 0.4 lower than it is today. Remember that the pH scale is logarithmic, so a change of 1 unit corresponds to a tenfold change in hydrogen ion concentration. Figure 8.6 shows the recent behavior of carbon dioxide mixing ratio, pressure of the same gas in the sea, and the average pH.

8.5.3 *Salinity*

About 85% of evaporation and 77% of precipitation occur over the oceans. The total amount of salt in the ocean does not change on a human time frame so that we can

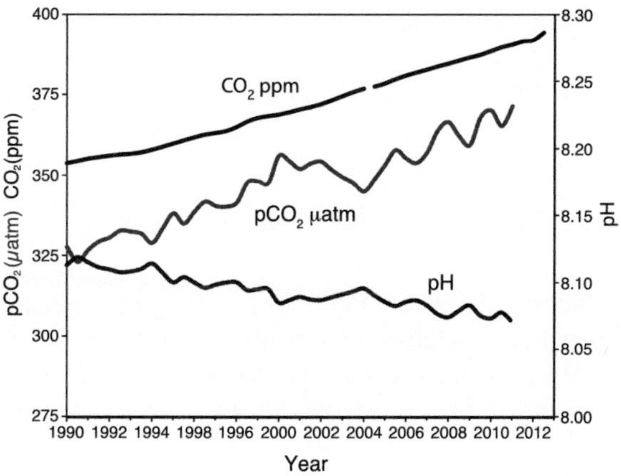

Fig. 8.6 The behavior of carbon dioxide pressure and acidity since 1990

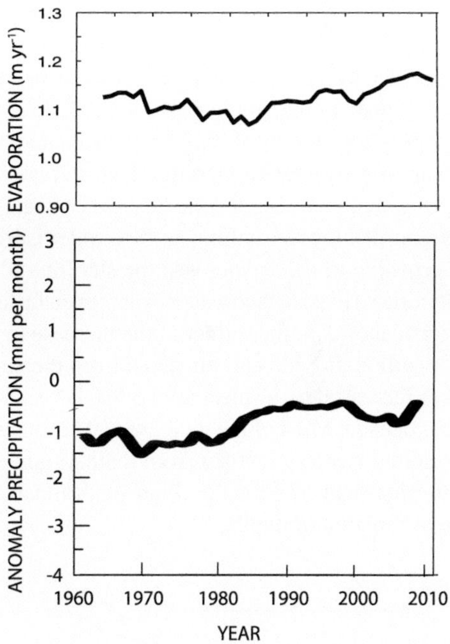

Fig. 8.7 The evaporation anomaly and precipitation in the last 50 years

only see variations in regions where evaporation dominates over precipitation (saltier in tropical regions) or vice versa (fresher or high-latitude regions). This distribution of salinity has been reinforced in recent years. In particular, the Atlantic Ocean where evaporation dominates has become saltier in the first 500 m while the Pacific Ocean has become fresher. Just because 77% of precipitation occurs over the ocean makes difficult to have a global picture of the effect on salinity because there are very few observing stations. Salinity can be affected by water runoff from the continents, and by the melting and freezing of sea ice or floating glacial ice. Freshwater added by melting ice on land will change global-averaged salinity, but changes to date are too small to observe. Figure 8.7 shows the anomaly in the precipitation and the evaporation in the last 50 years. It is quite evident the sparsity of data.

8.6 Static Stability of the Oceans

Quite recently a very interesting aspect of the modern ocean has been investigated related to the change of stability. As in the chase of the atmosphere the best way to define stability is through the definition of Brunt–Vaisala square of buoyancy frequency N^2 (see Appendix) that for the ocean reads

$$N^2 = -g\left(\alpha_T \frac{\partial \bar{T}}{\partial z} - \beta \frac{\partial \bar{S}}{\partial z} + \frac{g}{c_1^2}\right) \tag{8.29}$$

In order to have stability N^2 must be negative and the higher is the stratification the higher is the frequency. In (8.29), g is the acceleration of gravity; α_T and β, respectively, are the thermal expansion and the haline contraction coefficients; $\bar{T}$ and $\bar{S}$ are the average temperature and average salinity; and c_i is the sound speed. With the global warming the surface water heats up more rapidly than deep water while because of the ice melting surface water is fresher than deep water. Both these effects according to (8.29) increase the stability which results in a decrease in the vertical mixing so that surface waters warm even faster leading to, for example, more powerful hurricanes. Another effect of the increase in stability is the reduction in the transport of carbon dioxide and nutrients from the surface waters. The stratification globally has increased by a substantial 5.3% in recent decades (1960–2018) ; a rate of 0.90% per decade. Most of the increase (71%) occurred in the upper 200 m of the ocean and resulted largely (>90%) from temperature changes, although salinity changes play an important role locally. Figure 8.8 illustrate the trend and the percentage change as a function of depth.

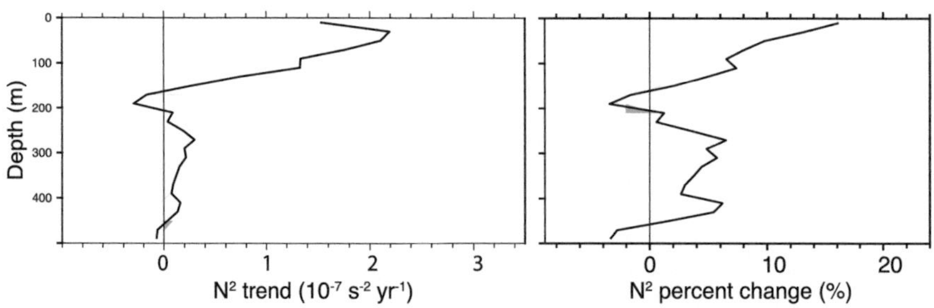

Fig. 8.8 The trend in N^2 and its percentage change of as a function of depth

Appendix

The Airborne Fraction

We refer to previous chapter and consider the reaction

$$CO_2(g) + CO_3^- \overset{H_2O}{\Longleftrightarrow} 2HCO_3 \tag{8.30}$$

With an equilibrium constant

$$K' = \frac{[HCO_3^-]^2}{P_{CO_2}[CO_3^{2-}]}$$

Then we ask how many dN mole of CO_2 added to the atmosphere remain there that is

$$f = \frac{dN_{CO_2(g)}}{dN} = \frac{dN_{CO_2(g)}}{dN_{CO_2(g)} + dN_{CO_2(aq)}} \tag{8.31}$$

where $dN_{CO_2(g)}$ and $dN_{CO_2(aq)}$ are, respectively, the added number of moles to the atmospheric and oceanic reservoirs at equilibrium. The definition of f must be made in infinitesimals term because the dissolution of carbon dioxide depends on the compound already present. Now we have

$$\frac{dN_{CO_2(g)}}{dP_{CO_2}} = \frac{N_a}{P} \tag{8.32}$$

where N_a are the total number of mole in the air $5.18 \times 10^{21}/29 \approx 1.8 \times 10^{20}$. We also have

$$dN_{CO_2(aq)} = V_{oc} dN_{CO_2(aq)} = V_{oc}(d[CO_2 \cdot H_2O] + d[HCO_3^-] + d[CO_3^{2-}]) \tag{8.33}$$

where V_{oc} is the ocean volume. Since the uptake of CO_2 by the ocean follow the equilibrium (8.30) we have $d[CO_2 \cdot H_2O[= 0]$ and also $d[HCO_3^-] = -2[CO_3^{2-}]$ so that (8.33) becomes

$$dN_{CO_2(aq)} = -V_{oc} d[CO_3^{2-}] \tag{8.34}$$

which substituted in (8.31) gives

$$f = \left(1 - \frac{V_{oc} P}{N_a} \frac{d[CO_3^{2-}]}{dP_{CO_2}}\right)^{-1} \tag{8.35}$$

We also have from equilibrium

$$[HCO_3^-]d[HCO_3^-] = K'(P_{CO_2}d[CO_3^{2-}] + [CO_3^{2-}]dP_{CO_2}) \tag{8.36}$$

Replacing $d[HCO3-] - 2d[CO3^{2-}]$ and $K' = K_H K_1/K_2$ into (8.36) we get

$$\frac{[CO_3^{2-}]}{P_{CO_2}} dP_{CO_2} = -\left(1 + \frac{4K_2}{[H^+]}\right) d[CO3^{2-}] \tag{8.37}$$

With the notation $\beta = 1 + 4k_2/[H^+] \approx 1.4$ we get from (8.35)

$$f = \left(1 - \frac{V_{oc} P K_H K_! K_2}{N_a \beta [H^+]^2}\right)^{-1} \tag{8.38}$$

Substituting the numerical values we get $f \approx 28\%$ which may increase to about 30% considering the mixing of the ocean layer but very far from the 0.50 we found before. The fraction missing is known as *missing sink* and could be attributed to the biosphere.

The Methane Oxidation

The process starts with a reaction with OH

$$\begin{aligned} CH_4 + OH &\rightarrow CH_3 + H_2O \\ CH_3 + O_2 + M &\rightarrow CH_3O_2 * M \end{aligned} \tag{8.39}$$

The methylperoxy radical CH_3OO initiates a process that results in ozone and carbon dioxide

$$\begin{aligned} CH_3O_2 + NO &\rightarrow CH_3O + NO_2 \\ CH_3O + O_2 &\rightarrow CH_2O + NO_2 \\ HO_2 + NO &\rightarrow NO_2 + OH \\ NO_2 + h\nu &\xrightarrow{O_2} NO + O_3 \\ CH_2O + OH &\xrightarrow{O_2} CO + HO_2 + CO \\ CO + OH &\xrightarrow{O_2} CO_2 + HO_2 \end{aligned} \tag{8.40}$$

Summing up

$$CH_4 + 8O_2 \rightarrow CO_2 + 4O_3 * 2H_2O \tag{8.41}$$

If the concentration of NO_2 is not high enough, ozone is destroyed by the odd hydrogen cycle. To understand a little better the importance of the nitrogen oxides, we may return to the previous set of reactions and in the case of low NO_2 the first reaction becomes so slow that the following process is favored:

$$\begin{aligned} CH_3O_2 + HO_2 &\rightarrow CH_3O_2H + HO_2 \\ CH_3O_2H + h\nu &\rightarrow CH_3O + OH \\ CH_3O + O_2 &\rightarrow CH_2O + HO_2 \end{aligned} \tag{8.42}$$

Also, in this case, in principle, formaldehyde is formed; however, the dissociation of CH_3O_2OH is so slow that it is more rapidly dissolved in water and then rained out. In this process, two radicals are lost (OH and normal upper HO_2), and carbon monoxide is not formed. More accurate calculations show that there is the loss of 3–4 radicals for poor NO_x air and a gain of 1–2 in situations of high mixing ratio.

The Stability of the Ocean Stratification

We assume density of water is a function of salinity S, temperature T, and pressure p(z) with z depth

$$\rho(z) = F(S, T, p(z)) \tag{8.43}$$

Then we have

$$\frac{1}{\rho}\frac{\partial \bar{\rho}}{\partial z} = \alpha_T \frac{\partial \bar{T}}{\partial z} + \beta \frac{\partial \bar{S}}{\partial z} + \gamma_T \frac{\partial \bar{p}}{\partial z} \tag{8.44}$$

where

$$\alpha_T = -\frac{1}{\rho}\frac{\partial \rho}{\partial T}, \qquad \beta = -\frac{1}{\rho}\frac{\partial \rho}{\partial S}, \qquad \gamma_T = -\frac{1}{\rho}\frac{\partial \rho}{\partial p} \tag{8.45}$$

At this point to derive the static stability conditions, we consider a parcel of water at depth z with density $\rho_0(z)$ and salinity S_0 and temperature T0. This parcel is raised to the depth $z + \delta z$ and let the parcel to expand at the lower pressure. The density will be

$$\begin{aligned} \rho_0(z + \delta z) = \rho_0(z) + \frac{\partial \rho}{\partial z}\delta z &= \rho_0(z) + \left(\frac{\partial \rho}{\partial p}\right)\left(\frac{\partial p}{\partial z}\right)\delta z \\ = \rho_0(z) + \left(\frac{1}{\rho_0}\frac{\partial \rho}{\partial p}\right)\left(\frac{\partial p}{\partial z}\right) &= \rho_0(z) - \gamma_a \rho_0^2 \delta z \end{aligned} \tag{8.46}$$

where $\gamma_a = (1/\rho_0)(\partial\rho/\partial p)$ is the adiabatic coefficient because we assume the parcel displacement is adiabatic. We can calculate now the acceleration of the parcel as $g\Delta\rho/\rho_0$ using (8.46)

$$\begin{aligned} g\frac{\Delta \rho}{\rho_0} = \frac{g}{\rho_0}[\bar{\rho}(z+\delta z) - \rho_0(z+\delta z)] =& \frac{g}{\rho_0}\left(\bar{\rho}(z) + \frac{\partial \bar{\rho}}{\partial z}\delta z - \rho_0(z) + \gamma_a \rho_0(z)^2 \delta z\right) \\ =& g\left(\frac{1}{\rho_o}\frac{\partial \rho}{\partial z} + \gamma_a \rho_0 g\right)\delta z = -N^2 \delta z \end{aligned} \tag{8.47}$$

That is the square of the frequency is proportional to δz. And using the derivative (8.44) we have

$$N^2 = -g\left(\frac{1}{\rho_o}\frac{\partial \rho}{\partial z} + \gamma_a \rho_0 g\right) = g\left[(\gamma_T - \gamma_a)\rho_0 g + \alpha_T \frac{\partial T}{\partial z} - \beta \frac{\partial S}{\partial z}\right] \tag{8.48}$$

Now the speed of sound is defined as

$$c_1^2 = \frac{1}{(\gamma_T - \gamma_a)\rho_0} \tag{8.49}$$

We have

$$N^2 = -g\left(\alpha_T \frac{\partial T}{\partial z} - \beta \frac{\partial S}{\partial z} + \frac{g}{c_1^2}\right). \tag{8.50}$$

References

AR5 Climate Change 2014: Impacts, Adaptation, and Vulnerability, Part B, Regional Aspect, Chapter 30,The Ocean (2014)

AR6 Climate Change 2022: Impacts, Adaptation and Vulnerability, Chapter 3, Ocean and Coastal Ecosystems and Their Services (2023)

Holland, H.D.: The Chemical Evolution of Atmosphere and Oceans. Princeton University Press (1984)

Jacob,D.: Introduction to Atmospheric Chemistry, 2nd edn. Princeton University Press (2025)

Li, G., Cheng, L., Zhu, J., et al.: Increasing ocean stratification over the past half-century. Nat. Clim. Chang. **10**, 1116 (2020)

Myhre, G., et al.: New estimates of radiative forcing due to well mixed greenhouse gases. Geophys. Res. Lett. **25**, 2715 (1998)

Pond, S., Pickard, G.L.: Introductory Dynamical Oceanography. Elsevier (1983)

Prather, M.J., Hsu, J.: Coupling of nitrous oxide and methane by global atmospheric chemistry. Science **330**, 952 (2010)

Tian, H., et al.: A comprehensive quantification of global nitrous oxide sources and sinks. Nature **586**, 249 (2020)

von Shuckman, K. et al.: Heat stored in the Earth system: where does the energy go? Earth Syst. Sci. Data **12**, 2013–2041 (2020)

Walker, C.G.: Evolution of the Atmosphere, Macmillan (1977)

Williams, R.G., et al.: How warming and steric sea level rise relate to cumulative carbon emissions. Geophys. Res. Lett. **39**, L19715 (2012)

Chapter 9
The Future of the Atmosphere and Oceans

Up to now we have narrated the history of the Earth's atmosphere and the oceans. It is a story that spans billions of years and it is time to speculate of what could happen in the future. It is already a very hard exercise to predict the near future, (let us say the next 1000 years) so that it looks as an academic endeavor to predict what could happen in the next few billion years. It is obvious that short-term changes are mainly the consequences of human activities while the long-term effect may depend on the expected evolution of our star (the Sun) or from some unexpected events like a repetition of the dinosaur impact. Even the human activities could bring the Earth system to much feared *tipping point* with a sudden and irreversible change in the environment or even materialize the *great filter* hypothesis, that is, the self-destruction of civilization.

9.1 Long-Term Warming

The effects of continuous emission of carbon dioxide have been reported by a number of studies (Ramanathan and Feng 2008; Matthews and Caldeira 2008; Allen and Frame 2009; Solomon et al. 2009) and they show that with 3.66×10^{12} tons of CO_2 (half of which has been already emitted) a warming of about $2\,^{\circ}C$ could be predicted with the simple result that warming is proportional to the total emitted gas. There is very illuminating contribution by Thomas Stocker published some year later on *Science* where it is assumed that emission of carbon dioxide follows a simple analytic rule

$$E(t) = \begin{cases} E_0 \cdot e^{r(t-t_0)} & t_0 < t < t_1 \\ E_0 \cdot e^{r(t-t_0)} \cdot e^{-s(t-t_0)} & t > t_1 \end{cases} \tag{9.1}$$

where E(t) is the anthropogenic emission, and t_0 and t_1 are, respectively, the starting time and the year of the emission reduction. The emission is at a rate r while the reduc-

G. Visconti, *Origin and Evolution of the Atmosphere and Ocean*,
https://doi.org/10.1007/978-3-032-02653-8_9

tion is at a rate s. We assume as in the paper $E_0 = 9.3GtCyear^{-1}(34GtCO_2year^{-1})$. The scenario path for $t > t_1$ in Eq. (9.1) implies that negative emissions (active removal of carbon from the atmosphere) on a global scale will not be realized anytime in the future. This should be considered as a conservative, but a realistic, assumption. The total cumulative emission at time t is given by the integral of (9.1)

$$C(t) = C_0 + \int_{t_0}^{t} E(t)dt = C_0 + E_0\left(\frac{1}{r} + \frac{1}{s}(1 - e^{-s(t-t_1)}\right)e^{r(t_1-t_0)} - \frac{E_0}{r} \quad (9.2)$$

where C_0 is the initial carbon content taken as 530 Gt. The final content C_∞ results then

$$C_\infty = C_0 + E_0\left(\frac{1}{r} + \frac{1}{s}\right)e^{r(t_1-t_0)} - \frac{E_0}{r} \quad (9.3)$$

Expression (9.2) can be easily converted into carbon dioxide partial pressure using the total mass of the atmosphere $\approx 5.1 \times 10^{18}$ kg

$$\frac{p_{CO_2}}{p_{atm}} = \frac{29}{12} \cdot \frac{1 \times 10^{12} \times C(t)}{5.1 \times 10^{18}}$$

Figure 9.1 gives an idea of the real emission as a function of time. Values used for r are 1.8% $year^{-1}$ while the removal rate is of the order $s = 3.\%year^{-1}$. An interesting result of this simple example is the calculation of the peak warming that is assumed to be proportional to the cumulative emission C_∞

$$\Delta T = \beta C_\infty = \beta\left[C_0 + E_0\left(\frac{1}{r} + \frac{1}{s}\right)e^{r(t_1-t_0)} - \frac{E_0}{r}\right] \quad (9.4)$$

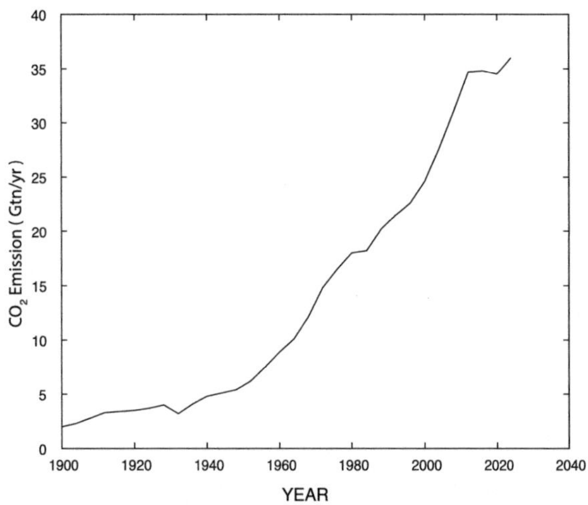

Fig. 9.1 The yearly emission of carbon dioxide from 1900 to 2024

While the minimum warming with the most aggressive reduction strategy $s \to \infty$ is

$$\Delta T_{min} = \beta \left[C_0 + \frac{E_0}{r}(e^{r(t_1-t_0)} - 1) \right] = \beta C_1 \tag{9.5}$$

where C_1 is the total emitted at time t_1. This approximation corresponds to cut to zero the emission after the time t_1. Then is very instructive to build a graph which gives the peak warming as in (9.4) as a function of the reduction rate s and the time t_1 of the beginning of emission reduction. A similar figure can be drawn for the case of ΔT_{min} when there is a cut to zero emission after t_1. As expected when the emission is reduced the amount of carbon dioxide continues to rise up to C_∞. These results are shown in Fig. 9.2. If we consider as an example an emission rate of 0.2% with a reduction of 0.3% starting from 2040 we got a $C_{infty} \approx 1500Gt$ that with a $\beta = 0.002°C\ Gt^{-1}$ gives a warming of about 3° which lasts for thousands of years. This warming would have consequences on the sea level that can be estimated to be 0.2–0.6 m per degree of warming which implies an increase up to about 2 m for the year 3000.

An important consequence of latest research is that the lifetime of CO_2 is much longer than the 300–400 years it was thought before. Montenegro et al. (2007) have shown that an impulse of 5000 Gt of 75% of carbon has a lifetime of about 1800 years while the remaining 25% has a lifetime of about 5000 years. Similar conclusions were reached by Archer et al. (1997). That means that even a total cut in the emission would imply much higher mixing ratio of carbon dioxide will persist in the atmosphere. If we indicate with CO_2^{peak} the peak concentration over its preindustrial value of CO_2^0, with $CO2^{equi}$ the 1,000-year quasi-equilibrium concentration, this can be expressed as

$$\begin{aligned} CO_2^{peak} &= CO_2^0 + AF^{peak} EMI \\ CO_2^{equi} &= CO_2^0 + AF^{equi} EMI \end{aligned} \tag{9.6}$$

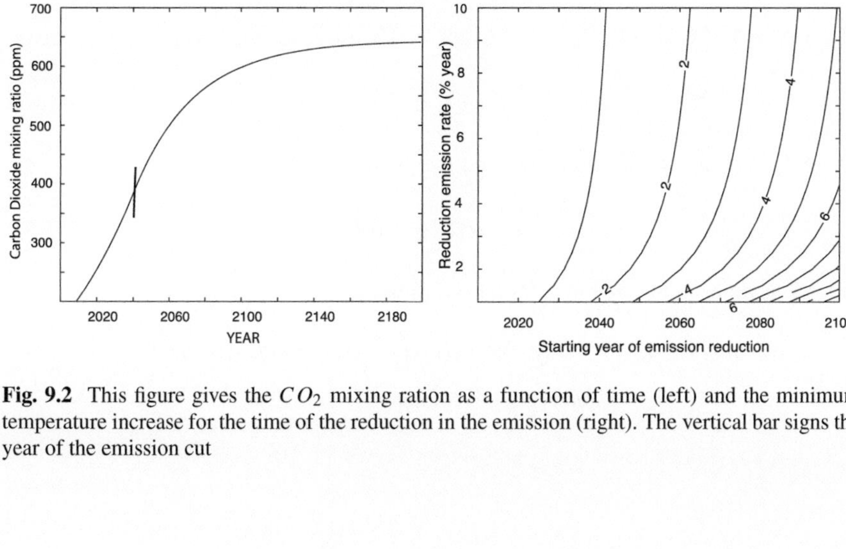

Fig. 9.2 This figure gives the CO_2 mixing ration as a function of time (left) and the minimum temperature increase for the time of the reduction in the emission (right). The vertical bar signs the year of the emission cut

where EMI is the total emission AF^{peak} and AF^{equi} the instantaneous (≈ 50) and equilibrium ($\approx 20\%$) airborne fraction, respectively. So we have

$$CO_2^{equi} - CO_2^0 = \frac{AF^{equi}}{AF^{peak}}(CO_2^{peak} - CO_2^0) \tag{9.7}$$

The factor AF^{equ} is largely governed by the Revelle factor (see Appendix) which regulates the long-term partitioning of carbon between the ocean and atmosphere/ biosphere system. Expression (9.7) means that if we assume $AF^{peak} \approx 50\%$ and $AF^{equi} \approx 20\%, CO_2^{peak} = 800ppm, CO_2^0 = 280ppm$ then the equilibrium concentration would be $\approx$40% above the preindustrial value even longer than 1000 years from now.

9.2 How Long the Biosphere Will Survive?

We have seen a simple way to have a long-term stabilization of climate through a negative feedback mechanism: as the luminosity of the Sun increases the temperature increases and so the precipitation which consumes carbon dioxide through erosion thus reducing the greenhouse effect. The decrease in carbon dioxide will affect negatively the biological productivity because the efficiency of the photosynthetic processes will decrease. The temperature could become so high that CO_2 could be depleted to the point as to inhibit photosynthesis completely. The result would be the complete cessation of biological activity that implies turning off the mechanism that regulates the temperature through erosion processes. Then we may ask based on such premises how long life on Earth could endure. This simple model devised by James Kasting and Ken Caldeira assumes that the solar luminosity changes as in paragraph (7.4)

$$S(t) = (1 - 0.38t/t_0)^{-1}S_0 \tag{9.8}$$

where $t_0 = 4.55Gyr$ and $S_0 = 1368\ wm^{-2}$. Then the surface temperature is given by

$$T = T_e + \Delta T \tag{9.9}$$

where T_e is related to the solar flux and the planetary albedo α

$$(1 - \alpha)S/4 = \sigma T_e^4 \tag{9.10}$$

and ΔT is determined by the CO_2 partial pressure

$$\begin{aligned}\Delta T =& 815.17 + (4.895 \cdot 10^{-7})T^{-2} - 3.9787 \cdot 10^5)T^{-1} \\ &- 6.7084\psi^{-2} + 73.23\psi^{-1} - 30.882\psi^{-1}T^{-1}\end{aligned} \tag{9.11}$$

where $\psi = \log P_{atm}$ with P_{atm} partial pressure of carbon dioxide. Expression (9.11) results from a fitting to a number of radiative-convective models. ΔT also depends on the temperature through the water vapor feedback, and the albedo depends on the temperature through the ice–albedo feedback

$$\alpha = 1.4891 - (6.5979 \cdot 10^{-3})T + (8.567 \cdot 10^{-7})T^2 \tag{9.12}$$

The erosion of siliceous rocks is a function of the ground water pH. A typical erosion reaction will include hydrolysis and the combination with carbon dioxide

$$Mg_2SiO_4 + 4CO_2 + 4H_2O \rightarrow 2Mg^{2+} + 4HCO_3^- + H_4SiO_4 \tag{9.13}$$

which is proportional to the activity of the hydrogen ions (α_{H^+}) and so the carbon dioxide pressure. Experimental data suggest for the erosion rate a dependence like

$$\frac{F_{wr}}{F_{wr,0}} = \left(\frac{P_{atm}}{P_{atm,0}}\right)^{0.3} exp\left(\frac{T - T_0}{13.7}\right) \tag{9.14}$$

With $T_0 = 15°$ C is the reference temperature. The biological activity in the soil has then a very important function in the weathering processes because it maintains in the soil a pressure that can be 30–40 times higher than the atmospheric pressure. For our model, it is very important to establish the dependence of the CO_2 pressure in the soil (P_{soil}) on the biological activity. The relationship we will use represents a very rough global average and is based on very sparse data. It is then necessary to use a good deal of imagination when using these data. Following Caldeira and Kasting, we can write

$$\frac{P_{soil}}{P_{soil,0}} = \frac{\Pi}{\Pi_0}\left(1 - \frac{P_{atm,0}}{P_{soil,0}}\right) + \frac{P_{atm}}{P_{soil,0}} \tag{9.15}$$

where Π is the biological productivity. The final link in the model must be between the productivity and the temperature because we see from Eq. (9.15) that if productivity falls below some threshold then the CO_2 pressure in the soil will become equal to the atmospheric pressure, and from that point, the photosynthetic activity will be reduced drastically. The link between productivity and temperature is given by

$$\frac{\Pi}{\Pi_{max}} = \left[1 - \left(\frac{T - 25}{25}\right)^2\right]\left[\frac{P_{atm} - P_{min}}{P_{1/2} + (P_{atm} - P_{min})}\right] \tag{9.16}$$

This relation points out that the productivity is controlled by pressure and temperature. If the temperature is higher than 50° C, then the productivity goes to zero. The same thing happens if the atmospheric pressure falls below a threshold (P_{min}) that, in our case, we have assumed as 10 ppm. Actually, this limit is reasonable for a particular kind of plants, called C_4 (e.g., sugarcane, corn), that utilize a photosynthetic process quite different from that of other plants (C_3). A few C_4 plants may survive even at lower pressure. In Eq. (9.16), $P_{1/2}$ is the pressure at which $\Pi_0 = \Pi$ when $T = T_0$ and $P_{atm} = P_{atm,0}$. It is a little surprising that the ocean apparently is not present in this model. Actually, things are a little different. Most of the photosynthetic activity take place in the ocean surface waters. Over the long term, the carbon dioxide added to the atmosphere by the volcanic activity must be balanced by the deposition of carbonates in the sediments. The weathering processes transport in the ocean calcium and magnesium ions that contribute to the alkalinity of the surface waters and consequently to the dissolution of carbon dioxide from the atmosphere. The model is usually solved for steady-state conditions using an iteration method. If we want to solve it as a function of time, we should take into account that the atmospheric CO_2 pressure is regulated on a global scale by volcanic emissions which are long-term source and destroyed by weathering processes. We have

$$\frac{dM_{CO_2}}{dt} = V - F_{wr} \tag{9.17}$$

where M_{CO2} is the carbon dioxide total mass per unit surface, V is the emission rate by the volcanic activity, and F_{wr} is the weathering rate. In order to use the relations written earlier, we multiply and divide through by the quantities with the subscript 0 to obtain

$$\frac{M_{CO_2,0}}{V_0}\frac{d}{dt}\left(\frac{M_{CO_2}}{M_{CO_2},0}\right) = \frac{V}{V_0} - \frac{F_{wr,0}}{V_0}\left(\frac{F_{wr}}{F_{wr,0}}\right) \tag{9.18}$$

The ratio between the present CO_2 mass ($5.71 \cdot 10^{16} mol$) and the emission rate from volcanoes ($5 \cdot 10^{12} mol year^{-1}$ gives a time constant of the order of 10 kyr. In the present condition, also the ratio of weathering/volcanoes is unity. We introduce a normalized time as

$$\bar{t} = tV_0/M_{CO_2,0}$$

So that (9.18) becomes

$$\frac{d}{dt}\left(\frac{M_{CO_2}}{M_{CO_2},0}\right) = \frac{V}{V_0} - \left(\frac{F_{wr}}{F_{wr,0}}\right) \tag{9.19}$$

which is the non-dimensional equation we may integrate. We make the assumption that the planet temperature will adapt instantaneously to the changing carbon dioxide partial pressure. In the original work by James Walker, the weathering rate depends on the partial pressure in the atmosphere because the idea is that the hotter the planet, the more abundant the precipitation. In the present work, the weathering rate depends on the partial pressure of carbon dioxide in the soil. This assumption ensures the interaction between the biosphere and the atmospheric pressure. We do not expect very much difference because if the partial pressure in the atmosphere and the soil is in equilibrium, then the ratios will not change too much. The results for this integration are given in Figs. 9.3 and 9.4. The first shows the partial pressure of CO_2 in the soil and the atmosphere (in ppm). As the solar constant increases (and then the

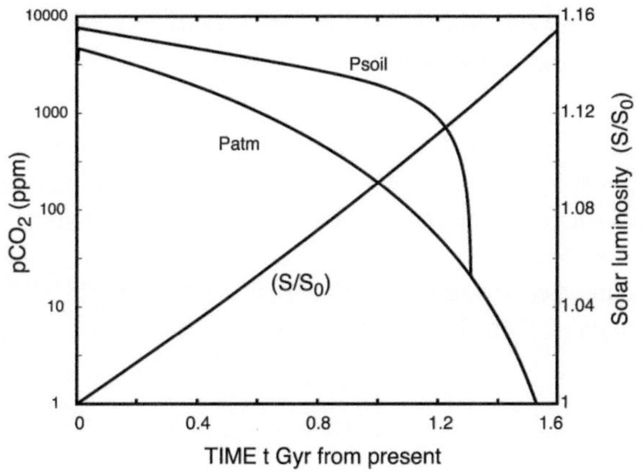

Fig. 9.3 The figure shows P_{soil} and P_{atm} of carbon dioxide as a function of the solar luminosity

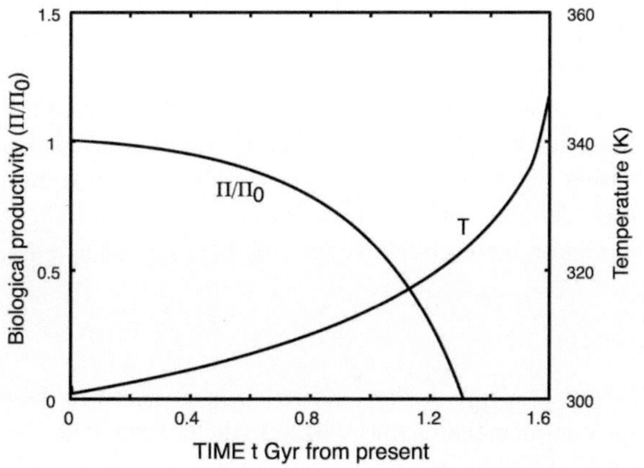

Fig. 9.4 The biological productivity Π and temperature as a function of time

temperature), both pressures decrease because on one side the biological productivity decreases and on the other the weathering increases. When the atmospheric pressure falls below 10 ppm, the photosynthetic activity ceases, and then we may assume the death of the biosphere. To interpret a little better these results, it is useful to examine the behavior of the biological productivity and temperature as shown in Fig. 9.3. The most evident result is a rapid temperature increase when the weathering or the biological activity is absent. One of the cases shown in the figure is when the atmospheric pressure is maintained constant at 320 ppm. The weathering has the effect of stabilizing the temperature because the reduction in the partial pressure of CO_2 compensates, with a lower greenhouse warming, for the increase in the solar constant. It is quite interesting in any case how the biological activity can stabilize more efficiently the temperature, at least in the short term, because it may accelerate the weathering rate through a higher pressure in the soil.

9.3 The Academic Lifespan of the Biosphere: Longer or Shorter

The work of Caldeira and Kasting was written as an improvement on a previous paper by Lovelock and Whitfield as proof of the esoteric Gaia hypothesis. Following this nice self-contained work, a number of variations on the same theme have been developed since then as typical behavior of the academic people. Timothy Lenton published a work where the erosion rate was taken as a function also of the productivity so that (9.14) was modified as

$$\frac{F_{wr}}{F_{wr,0}} = \left[\left(1 - \frac{1}{\alpha}\right)\frac{\Pi}{\Pi_0} + \frac{1}{\alpha}\right]\left(\frac{P_{atm}}{P_{atm,0}}\right) exp\left(\frac{T - T_0}{13.7}\right) \tag{9.20}$$

where α represents the amplification due to processes other than enhanced soil pCO_2. The case $\alpha = 1$ reduces to case we discussed before while the author uses a value around 20 that would enhance considerably the erosion lowering then the temperature and extending the lifespan of the biosphere up to a little more than a Gyr against the value of 0.8 Gyr obtained by Caldeira and Kasting. In a previous work of a year earlier Siegfried Franck and coworkers observed that the plate tectonics should be considered as contribution to the carbon dioxide budget. Consequently they proposed a relation between the weathering rate and the spreading rate of the plates

$$f_{wr} \dot{f}_A = f_{sr} \tag{9.21}$$

where $f_{wr}/f_{wr,0}$ is the weathering normalized to the present value, $f_A = A_c/A_0$ is the continental area normalized to the present value, and $f_{sr} = S/S_0$ is the spreading rate normalized by the present value. The ratio f_{sr}/f_A is called geophysical forcing ratio and decreases with time so that when is taken into account in Eq. (9.17) results in a shortening of the productivity of about 300 million years from about 1 Gyr to less than 0.7 Gyr. This is nothing serious considering that the number is related to large uncertainties in the spreading rate and the variation in the continental area. The more recent variation of the theme is a work of Graham et al. where they include in the calculation of productivity also the C_3 plants. As we have seen before these kind of plants may starve earlier for the lack of carbon dioxide. They assume the weathering rate equal the outgassing so that (9.14) is written as

$$\frac{V}{V_0} = \left(\frac{P_{soil}}{P_{soil,0}}\right)^{\beta} \exp\left(\frac{\Delta T}{T_d}\right) \tag{9.22}$$

Taking now $V = V_0$ and the logarithm we have

$$T_d \beta = -\frac{\Delta T}{\log\left(\frac{P_{soil}}{P_{soil,0}}\right)} \tag{9.23}$$

So the trajectories are determined by the product of β and T_d, i.e., simulations with $T_d = 20K$ and $\beta = 0.25$ produce the same trajectories as simulations with $T_d = 10K$ and $\beta = 0.5$. They use a number of combinations and, in particular, $T_d = 31K$ and $\beta = 0.41$ that were used by Krissansen-Totton (KT) and Catling to obtain for the C_4 plant a much longer survival up to about 1.8 Gyr. It has to be noticed that the data supporting the KT hypothesis extend to only 100 Myrs ago. Except for some additional merits for their list of publications these efforts do not add very much to the conclusion drawn by Caldeira and Kasting paper: life on Earth will end sometime one billion year from now.

9.4 Oxygen in the Atmosphere May Disappear

At this point we ask what would be the consequences of the decrease in the amount of carbon dioxide in the atmosphere predicted by the Caldeira and Kasting paper. The lower amount of CO_2 will affect directly the oxygen in the atmosphere. As a matter of fact as we have seen in Chap. 6 oxygen is produced in the photosynthetic process

$$CO_2 + H_2O \xrightarrow{light} CH_2O + O_2 \tag{9.24}$$

At the bottom of the ocean mud the organic carbon (CH_20) is converted to CO_2 by using Fe_2O_3 and H_2SO_4 as oxygen sources, i.e., without consuming O_2. In this manner, the O_2 previously generated in the production of the organic carbon is not consumed in the eventual oxidation of this organic carbon, resulting in a net O_2 source

$$Fe_2O_3 + 8H_2SO_4 + 15CH_2O \rightarrow 4FeS_2 + 15CO_2 + 23H_2O \tag{9.25}$$

which coupled with photosynthesis

$$15CO_2 + 15H_2O \xrightarrow{light} 15CH_2O + 15O_2 \tag{9.26}$$

gives

$$2Fe_2O_3 + 8H_2SO_4 \rightarrow 4Fe_2S + 8H_2O + 15O_2 \tag{9.27}$$

The net result is that atmospheric oxygen is produced in the formation of organic material and this source is compensated when the deposits of pyrite are brought to surface and re-oxidized as in (9.27). This is also a stabilizing mechanism: should the weathering consume more oxygen this would increase the burial of organic material, which would increase the source of O_2 and act as a negative feedback. Kazumi Ozaki and Christopher Reinhard have shown using a quite sophisticated procedure that the reduction of carbon dioxide could bring to a depletion of atmospheric oxygen down to less than 0.01 PAL in about 1 Gyr time. Again this is an academic exercise because the destiny of the Earth could be much more dramatic.

9.5 Some Future of the Earth

Up to now we have considered the way the climate system behaves while we all know even from the newspapers what could be our near future. Climate change is going to influence the life on Earth and possibly even the survival of our species on this planet. Axel Kleidon and Adam Frank among others have elaborated a simple model on our future that is worth some mention. Kleidon has studied for several years how the human activity may change the famous NPP (net primary production) and he is among the inventors of the so-called *human appropriation of net primary production, HANPP*: This corresponds to the fact that human activity may subtract carbon to NPP by agricultural and other activities. As percentage of HANPP grows the NPP decreases because there is less biomass available. In some measure then the human activity may change the conditions of survival of the species. The model for the population N assumes a basic rate of growth R_0

$$\frac{1}{N}\frac{dN}{dt} = A_0 - B_0 \tag{9.28}$$

where A_0 and B_0 are the rate of birth and death, respectively. This grow rate may increase due to the energy harvesting which is related to the carbon dioxide increase

$$R_+ = R_0\left(1 + \frac{P - P_0}{\Delta P}\right) \tag{9.29}$$

where P is the partial pressure of CO_2 and ΔP is the change required to double the growth rate. On the other hand, the increased growth rate increases the temperature and this feedback on the growth rate itself

$$R_- = R_0\left(\frac{T - T_0}{\Delta T}\right)^2 \tag{9.30}$$

where ΔT is called the *temperature tolerance* that is the maximum temperature increase that an individual could tolerate. We have an equation for the population growth

$$\frac{dN}{dt} = min[NR_+, R_0(N_{max} - N)] - NR_- \tag{9.31}$$

In this case, N_{max} corresponds to the carrying capacity of the Earth, a concept familiar with the logistic growth model

$$\frac{dN}{dt} = NR\left(1 - \frac{N}{N_{max}}\right) \tag{9.32}$$

The production P of CO_2 is modeled as proportional to N

$$\frac{dP}{dt} = CN \tag{9.33}$$

As for the climatic part the model assumes following what we have seen before that the temperature

$$T \approx T_0 + \Delta T_F \log \frac{P}{P_0} \tag{9.34}$$

where ΔT_F is the temperature change necessary for the sensitivity to drop by a factor e. Now we can write the change in temperature with time

$$\frac{dT}{dt} = CN\frac{dT}{dP} = CN\frac{\Delta T_F}{P} = CNDe^{-\frac{T-T_0}{\Delta T_F}} \tag{9.35}$$

where $D = \Delta T_F / P_0$ while from (9.69) to (9.71) we have

$$\frac{dN}{dt} = min\left\{R_0 N\left[1 + \frac{P_0}{\Delta P}\left(e^{\frac{T-T_0}{\Delta T_F}} - 1\right)\right], R_0(N_{max} - N)\right\} - R_0 N\left(\frac{T - T_0}{\Delta T}\right)^2 \tag{9.36}$$

It is now convenient to normalize this equation dividing by $R_0 N_{max}$ and (9.75) by $R_0 \Delta T$, so that we define the quantities. η is defined as the *normalized population* N/N_{max}, τ is defined as the *normalized time* $\tau = R_0 t$ while the *normalized temperature* ϵ is defined as $(T - T_0)/\Delta T$. We also define some additional variables. First of all the *timescale for population growth* $t_G = 1/R_0$, the *timescale for technological advancements* derived from (9.73) as $t_T = \Delta P / C N_{max}$ and then the *timescale for climate change* defined as $t_C = \Delta T / C N_{max} D$. Now the ratio $\theta = t_C / t_T = \Delta T / \Delta P D$ can be interpreted as the expected change in population growth rate as the consumption of fossil fuel is such that the temperature change by ΔT. After the substitutions are made we get two equations

$$\begin{aligned}\frac{d\eta}{d\tau} &= min\left\{\eta\left[1 + \frac{\theta}{\alpha}(e^{\alpha.\epsilon} - 1)\right], 1 - \eta\right\} - \eta\epsilon^2 \\ \frac{d\epsilon}{d\tau} &= \gamma\eta e^{-\alpha\epsilon}\end{aligned} \tag{9.37}$$

where $\alpha = \Delta T / \Delta T_F$ is the ratio of temperature change that affects biology to temperature change that affects climate sensitivity, and $\gamma = DCN_{max}/R_0\Delta T$ is the *normalized forcing*. Integration of these equations is made in the appendix and we show here two typical results. Figure 9.5 shows the behavior of the population η and temperature ϵ. Our best Earth model had a $\gamma = 1.94$, $\theta = 14.45$, and $\alpha = 1.52$ and these values are obtained fitting the historical record. This figure illustrates the case

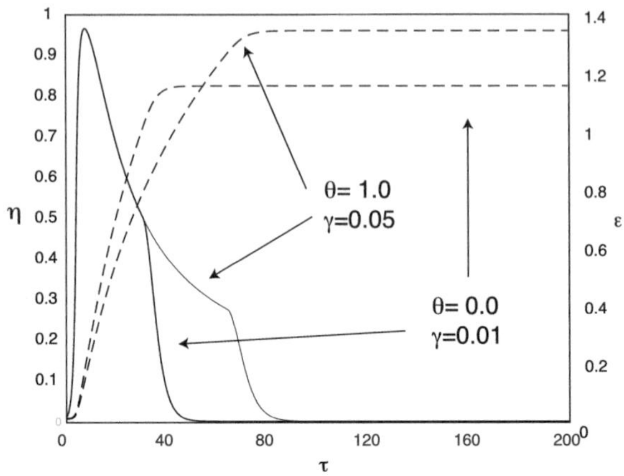

Fig. 9.5 The normalized population (solid line) and normalized temperature (dashed lines) for two values of θ and the normalized forcing γ

in which the forcing is weak from 0.01 to 0.05 and θ is between 0 and 1. The general feature of these solution is the rapid increases of population up to the maximum capacity ($\eta = 1$) followed by a decline with different slope. For $\theta = 0$ (no help from technology) after $\eta = 0.5$ the decline is accelerated while this behavior is delayed for $\theta \neq 0$ or very weak forcing. On the other hand, the temperature will raise within the population up to the point the same starts to decline. The stabilization of temperature is roughly regulated either by θ or the entity of forcing. At this point, it is interesting to define the anthropocene in these results. We can assume that we enter the anthropocene when the population growth is strongly influenced by the increase in temperature. This means when the population is close to the carrying capacity. This very simple model has been described just for fun.

9.6 The Fate of the Oceans

It is quite hard to find references about the future state of the oceans. We refer to a paper by Christine Bounama and coworkers that since the beginning affirm that terrestrial oceans are bound to disappear. They consider that *transport of water to the surface is five times smaller than water movement to the mantle, so that the Earth will lose all its seawater in one billion years from now*. This is one way in which the ocean may disappear. The other way is through the runaway greenhouse triggered by the increase in surface temperature following the increase of the solar output from the Sun. We have discussed of the runaway greenhouse already in Chap. 5 but referring mainly to the so-called *Komabayashi–Ingersoll limit* which considers a moist troposphere in contact with the ocean. To calculate a radiation limit, we assume that it is saturated with water vapor and this determines the water content of the region above (the stratosphere) which is radiative equilibrium. With these assumptions the optical thickness is directly related to the water content

$$\tau = k \frac{r e_s(T) m_v}{m g} \tag{9.38}$$

with k absorption coefficient, r is relative humidity, e_s is saturation pressure, and m_v and m are molecular masses of vapor and atmosphere. The saturation pressure is determined by its value at the tropopause so that for a constant mixing ratio the optical thickness in the stratosphere is determined by the temperature at the tropopause. The optical thickness will be determined only by the temperature and the radiative limit will be given when the radiative equilibrium temperature is tangent to this curve (Fig. 5.7). In this case, the Komabayashi–Ingersoll limit is given by about 385 wm^{-2}. However as the temperature increases more water is evaporated from the ocean increasing the water pressure up to the background atmosphere (10^5 Pa) to a mixing ratio 0.5. At that point, the temperature makes the ocean to boil but because of the growing pressure this will happen at the critical point of water (647 K) when the saturation pressure will be roughly 260 bar. However adding water vapor will not change the temperature structure and effective emission to space takes place from $\tau \approx 1$. The radiative flux to space is decoupled from surface temperature, with the radiation limit

$$F_{out} \approx \sigma T_{trop}^4 \tag{9.39}$$

This as the moist troposphere limit or Simpson–Nakajima limit. Actually detailed calculations show that this limit whose value is around 385 wm^{-2} depends on the atmospheric content of other gases like carbon dioxide and has been recently further refined by Goldblatt et al. to 285 wm^{-2}. However this would correspond to the evaporation of the ocean whose mass (1.4×10^{21} kg) would correspond to a pressure of

$$\frac{1.4 \times 10^{21} \times 9.8}{5.1 \times 10^{14}} \approx 270 bar$$

Much heavier than the Venus atmosphere. To evaporate all this water you need

$$1.4 \times 10^{21} \times 2.5 \times 10^{6} \approx 3.5 \times 10^{27} joule$$

Assuming an absorbed flux of 200 wm^{-2} it would take

$$\frac{3.5 \times 10^{27}}{200 \times 5.1 \times 10^{14}} \approx 1000 yrs$$

Which apparently is much shorter than reported by Abe et al. (2011). The evaporation of the ocean as a consequence of a runaway is one of the possible causes for the disappearance of the ocean. The other is related to the water budget in the mantle of the Earth. Different estimates show that the transport of water to the surface is five times less (2×10^{11}kg yr^{-1}) than the transport to the mantle (1×10^{12}kg yr^{-1}). This means that in a time frame of 1 billion year the ocean will lose about one-third of its water. Bounama et al. have presented a simple model with elements shown in

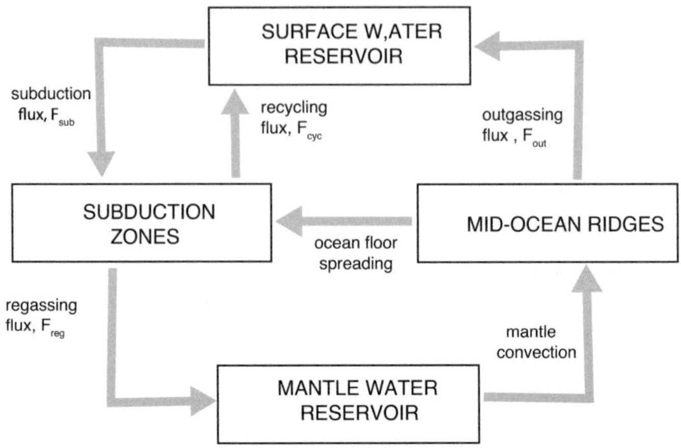

Fig. 9.6 The block diagram to calculate the water exchange between the surface and the mantle

Fig. 9.6 and the exchange of water is such that we write for the surface water reservoir

$$\frac{dM_{surf}}{dt} = F_{out} - F_{reg} \tag{9.40}$$

where F_{out} is the outgassing flux from the mantle and F_{reg} is the regassing flux. These fluxes will be a function of the spreading rate of the ocean floor

$$S_r = q_m^2 \frac{\pi \kappa A_0}{4k^2 (T_m - T_s)^2} \tag{9.41}$$

where q_m is the convective flux, κ is the thermal diffusivity, and k is the thermal conductivity. A_0 is the difference between the surface of the Earth and that of the continent while T_m and T_s are the temperature of the mantle and the surface, respectively. The outgassing flux can be written as

$$F_{out} = \rho_w d_w f_w S_r \tag{9.42}$$

With ρ_w density of water in the mantle, d_w is the depth of the ascending mantle, and f_w is the fraction of degassing water. In a similar way, the regassing flux is

$$F_{reg} = \rho_{bas} f_{bas} d_{bas} S_r R_{H_2O} \tag{9.43}$$

where ρ_{bas} is the density of the basalt layer, f_{bas} is the fraction of water in the basalt layer, d_{bas} is thickness, and R_{H_2O} is the regassing ratio as the ratio between the regassing flux and the total flux. This is taken as

$$R_{H_2O} = m[T_m(0) - T_m(t)] + R_{H_2O}(0) \tag{9.44}$$

On the other hand, ρ_w is given by

$$\rho_w = \frac{M_{tot} - M_{surf}}{V_m} \tag{9.45}$$

where M_{tot} is the total mass in the Earth and V_m is the volume of the mantle. The only complicated stuff in this model is the calculation of the convective flux but in our case we have made a drastic simplification by adopting the parameterization of Davies already illustrated in Chap. 2

$$\frac{q_m}{q_0} = \left(\frac{T}{T_0}\right)^m \tag{9.46}$$

that can be re-normalized at the present values. Then the temperature is regulated by the non-dimensional equation

$$\frac{dT}{dt} = h - T^m \tag{9.47}$$

where to include the radioactive heat source we use

$$h = h_0 e^{-t/\lambda} \tag{9.48}$$

With λ some normalized decay constant. Because $T_m \gg T_s$ we just assume that $T_m - T_s \approx T_m$. In this simple case, the equations for the surface water, M_{surf} and the temperature of the mantle are completely decoupled so that, in principle, the temperature of the mantle is determined by

$$\frac{dT_m}{dt} = \frac{Q}{\rho c} - \frac{3R_m^2 q_m}{(R_m^3 - R_c^3)} \tag{9.49}$$

where ρ is the density, Q is the heath generation by the radioactive sources, c is the specific heat at constant pressure, q_m is the heat flow from the mantle, and R_m and R_c are the radius of the mantle and of the core, respectively. Relation (9.49) is very similar to (9.47) and should give a simple decrease temperature in time that substituted in (9.40) gives also the solution for the mass of the surface water which decrease with time. The solution found by Bounama et al. gives a halved ocean mass in about a billion year from now.

9.7 Predicting Catastrophes: Tipping Points and the Great Filter

We were in doubts whether to initiate another chapter or open a long paragraph on this interesting and rather new perspective on the future of the Earth environment. We

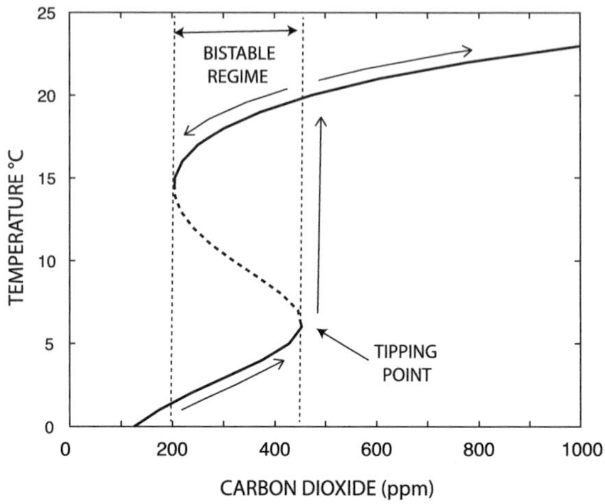

Fig. 9.7 The hysteresis cycle and tipping point for an energy balance model with varying CO_2. The dashed line indicates the unstable regime

had already a glimpse of the tipping point when talking about the Budyko model in Chap. 7. We saw that decreasing the solar constant the ice line would plunge from mid-latitudes to the equator. This jump was accompanied by the hysteresis phenomena that is the system would return to an ice-free planet on a different path because the albedo of the ice-covered Earth was much higher and would require a large increase in the value of the solar constant (see also Fig. 9.11). The *tipping point* can be defined as a threshold that once reached makes the system change its state. This is shown also for a large increase in CO_2 as in Fig. 9.7 and discussed later. Before going to some detail we can talk about *the great filter*. This concept was introduced in 1998 by the economist Robert Hanson as one way to explain the apparent absence of any signal that would reveal extraterrestrial civilization. This absence would be explained because before any civilization reaches a technological capacity to communicate it would self-destroy through a global warfare or an environmental catastrophe. As we will see tipping points could explain some long-term climate change (i.e., glaciations) while the great filter could imply some very destructive change.

9.7.1 *The Stommel Tipping Point*

The Stommel two box model offers a classic example of an abrupt change in the circulation at a certain level of freshwater forcing by transitioning from a stable state to another, and the threshold is mathematically identified as a *bifurcation point*. Figure 9.8 illustrates the reduction of the entire ocean to two boxes, one at equatorial latitudes with temperature T_1 and salinity S_1 and the other at northern latitude with temperature T_2 and salinity S_2. Between these two boxes the water is transported by the density differences

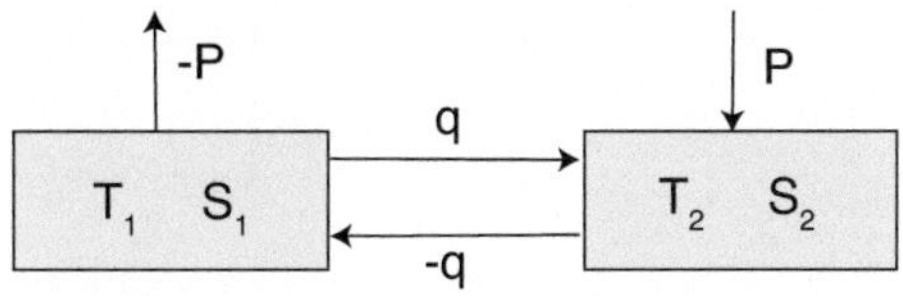

Fig. 9.8 The two box Stommel model

$$q = k(\rho_2 - \rho_1) = k[\alpha(T_2 - T_1) + \beta(S_2 - S_1)] \tag{9.50}$$

where the equation of state has been used

$$\rho(T, S) = \rho_0[1 + \alpha(T - T_0) + \beta(S - S_0)] \tag{9.51}$$

where k is a hydraulic constant; a and b are the thermal and haline expansion coefficients, respectively. If q>0 there is a poleward surface flow because high-latitude density is greater than low-latitude density, and vice versa, temperature difference dominates the density difference and drives the circulation whereas the salinity difference brakes it, and vice versa. We also assume there is a freshwater flux P that leaves the equatorial box (net evaporation) and enter the northern box (net precipitation), so the salinity balance reads

$$\begin{aligned} \frac{dS_1}{dt} &= P + |q|(S_2 - S_1) = P + |q|\Delta S \\ \frac{dS_2}{dt} &= -P - |q|(S_2 - S_1) = -P - |q|\Delta S \end{aligned} \tag{9.52}$$

Subtracting the second from the first we get at the steady state

$$P + |q|\Delta S = 0$$

Substituting for q and solving for ΔS we get

$$\Delta S = \begin{cases} -\frac{\alpha\Delta T}{2\beta} \pm \sqrt{\left(\frac{\alpha\Delta T}{2\beta}\right)^2 - \frac{P}{k\beta}}, & \text{if } q > 0 \\ -\frac{\alpha\Delta T}{2\beta} + \sqrt{\left(\frac{\alpha\Delta T}{2\beta}\right)^2 + \frac{P}{k\beta}}, & \text{if } q < 0. \end{cases} \tag{9.53}$$

We introduce the non-dimensional quantities

$$\delta = -\frac{\beta\Delta S}{\alpha\Delta T}, \qquad E = \frac{\beta P}{k(\alpha\Delta T)^2} \tag{9.54}$$

so that (9.53) reduces to

$$\delta = \begin{cases} \frac{1}{2} \pm \sqrt{\frac{1}{4} - E}, & \text{if } q > 0 \\ \frac{1}{2} + \sqrt{\frac{1}{4} + E}, & \text{if } q < 0 \end{cases} \tag{9.55}$$

We now can study the stability of solutions. To this end we write (9.52) as

$$\frac{d}{dt}\Delta S = -2P - 2|q|\Delta S = -2P - 2(k\alpha\Delta T - k\beta\Delta S)\Delta S \tag{9.56}$$

And perturb the steady-state solution q, $\Delta\bar{S}$ as

$$\Delta S = \Delta\bar{S} + \Delta S', \qquad q = \bar{q} + q'$$

And (9.56) becomes

$$\frac{dS'}{dt} = -2|q|S' \pm k\beta S'(\Delta\bar{S} + S') \;\; + : \mathrm{q} > 0 \;\; - : \mathrm{q} < 0 \tag{9.57}$$

This expression linearized gives assuming that $S' = S_0 e^{\lambda t}$

$$\lambda = -2|\bar{q}| \pm 2k\beta\Delta\bar{S} = -2k|\alpha\Delta T - \beta\Delta\bar{S}| \pm 2k\beta\Delta\bar{S} \tag{9.58}$$

In the case q>0 (northward flux) the instability condition ($\lambda > 0$) reads $2\beta\Delta S > \alpha\Delta T$ that is $\delta > 1/2$. Notice also that the normalized flux

$$\bar{q} = \frac{q}{k\alpha\Delta T} = \frac{k(\alpha\Delta T - \beta\Delta S)}{k\alpha\Delta T} = 1 - \delta \tag{9.59}$$

So that for positive fluxes δ must be comprised between 0.5 and 1 in order to have instability. On the other end when q<0 we have stability for $\delta > 1$. Based on (9.55)

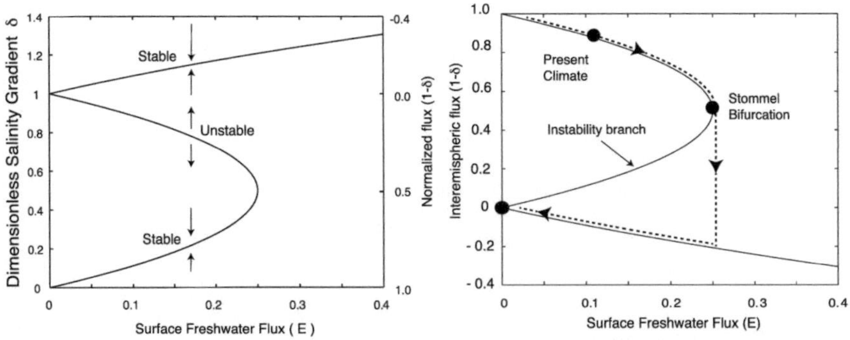

Fig. 9.9 The solution to the Stommel box model. On the left are shown the regions of instability while on the right the hysteresis cycle is shown

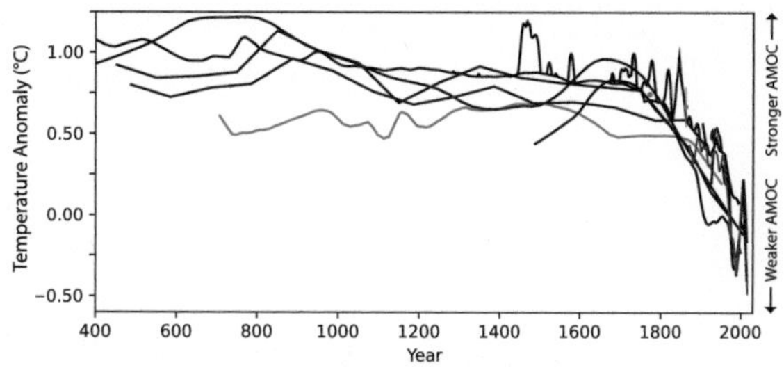

Fig. 9.10 The temperature anomaly of the oceanic "cold trap" in the last 1600 years. Also indicated is the intensity of the AMOC

the situation can be summarized as in Fig. 9.9 where we show the solution of the Stommel model as a function of δ and the freshwater flux E. For all values of E and $0.5 < \delta < 1$ we have instability while for $0 < \delta < 0.5$ we have a *direct* northward current and for $\delta > 1$ we have *indirect*. In the direct circulation, heat is transported from low latitudes to high latitudes while the salinity is transported from low to high latitudes while is the atmosphere to transport in the same direction freshwaters (evaporation at the equator and precipitation at the pole). The more salty water sinks where it is cold. In the indirect circulation, freshwater is still transported in the atmosphere while in the surface ocean the current is southward ($\delta > 1$). This means a much larger transport of heat in the atmosphere and the salt being moved at depth between equator and the pole and water sinks where it is warm. These results from a simple model can be transported to the real ocean and bring to an hysteresis behavior as shown in Fig. 9.9. If the freshwater flux increase (for example, by melting of the ice sheets), the salinity decreases and the climate moves from point A to the bifurcation point B where following the instability jumps to point C. If from there the freshwater flux is reduced there is not an immediate return to the bifurcation point and the circulation is now southward ($1 - \delta < 0$. What is the connection with reality? Stefan Rahmstorf is a great supporter of this view and he argues that in the last 2000 years the Atlantic Meridional Ocean Circulation (AMOC) has decreased its intensity as shown in Fig. 9.10. This could be the trigger for a new ice age (The Day After Tomorrow Movie) and so represents a practical consequence of tipping point.

9.7.2 *Tipping Points in the Climate*

We mentioned already that Energy Balance Climate Model do hint to the existence of tipping point. That is related to a slight change in the solar constant. The behavior of the Budyko-type model result is more clear if we formulate it in the following way after Zalapin and Ghil. They assume an albedo function of the temperature as

$$\alpha(\kappa, T) = c_1 + 0.5c_2((1 - \tanh(\kappa(T - T_c))) \tag{9.60}$$

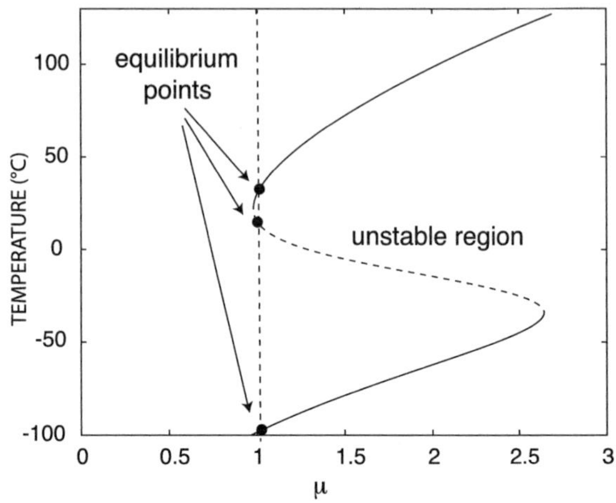

Fig. 9.11 The ice line as a function of the solar constant

where $c_1 = 0.15$ and $c_2 = 0.7$ and $T_c = 260K$. The emissivity is of the form

$$\epsilon = 1 - m \tanh((T/T_0)^6) \tag{9.61}$$

where m=0.4 and $T_0 \approx 284K$. The solar constant Q_0 is multiplied by a factor μ so that at the equilibrium either by comparing the emitted radiation $\epsilon\sigma T^4$ to the absorbed ones

$$\mu Q_0(1 - \alpha(\kappa, T) = \epsilon\sigma T^4 \tag{9.62}$$

These simple relationships can be used to study the equilibrium point of the climate either by comparing the emitted radiation $\epsilon\sigma T^4$ to the absorbed ones $\mu Q_0(1 - \alpha(\kappa, T)$ or by expressing the temperature as a function of μQ_0. This is shown in Fig. 9.11 where it results that for $\mu = 1$ there are three equilibrium points two of which are stable (the present climate and the snowball Earth) and the third unstable because lies in the region where the temperature decreases for an increase of solar constant, it looks like that this tipping point has been hit a few times in the transition between the present climate and the snowball Earth. There are several other elements besides the ice feedback that could trigger a tipping point (see Armstrong et al. 2022) but not of that could be proved. Similar question was posed by Steffen et al. (2019). However a similar tipping point can be found also for the greenhouse forcing and essentially based on the albedo dependence on temperature. We may refer to different form which hypothesize a drastic decrease around 10° theta can be written as

$$\alpha(T) = \alpha_1(1 - \Sigma T) + \alpha_2 \Sigma T \tag{9.63}$$

where $\alpha_1 = 0.52$ and $\alpha_2 = 0.47$ and

$$\Sigma T = \frac{T - T_1}{T_2 - T_1} H(T - T_1)H(T_2 - T) + H(T - T_2)$$
$$H(T) = 0.5(1 + \tanh(T/T_\alpha)) \quad (9.64)$$

where $T_\alpha =$. Then from equilibrium between absorption and emission we get for the partial pressure of carbon dioxide

$$C = C_0 exp\left[\frac{\epsilon(T)\sigma T^4 - Q_0(1 - \alpha(T))}{A}\right] \quad (9.65)$$

where the emissivity ϵ is a decreasing function of T while A is part of the forcing term $A\log(C/C_0)$ with C_0 preindustrial partial pressure of CO_2 taken as 280 ppm. Using the parameter given by Ashwin et al. (2020) we can draw Fig. 9.7 where above some level of carbon dioxide the temperature jumps of about 15 °C. For this jump we have no proofs that could have happened in the past.

9.8 Can We Forecast Tipping Points?

This huge catastrophe apparently can be predicted as affirmed by several papers. The starting point is to assimilate the tipping point to a basin of attraction as shown in Fig. 9.12. As shown in the appendix this can be traced down to the potential function used in the zero-dimensional energy balance model. The basin of attraction can be of high or low resilience. In the first case, we have steep walls and a quite rapid recovery rate from random perturbations. In the case of low resilience and the system is near the transition the basin of attraction shrinks the resilience is lower and the rate of recovery from small perturbations is lower. The consequence is in the first case a lack of correlation between subsequent states which are more evident in the second case. In the appendix, we give an analytical proof of such behavior. There is a beautiful paper by Vasilis Dakos et al. on PNAS which present several examples taken from the Earth climatic history.

9.8.1 *The Great Filter*

Another possible traumatic development in the Earth history (and so its fluid component) has been invented in 1998 by the economist Robin Hanson and it is known as the *The Great Filter* hypothesis. Hanson has developed this idea from the fact that we humans seem to be the only inhabitants of the universe. Considering this as highly unlikely Hanson argues that in the phases of development of any form of life there exist critical steps which he characterizes as great filters. These could include the abiogenesis itself (that is the origin of life from inorganic matter), or the transition from *prokaryotes* (the most basic type of single-cell organism) to *eukaryotes* (a

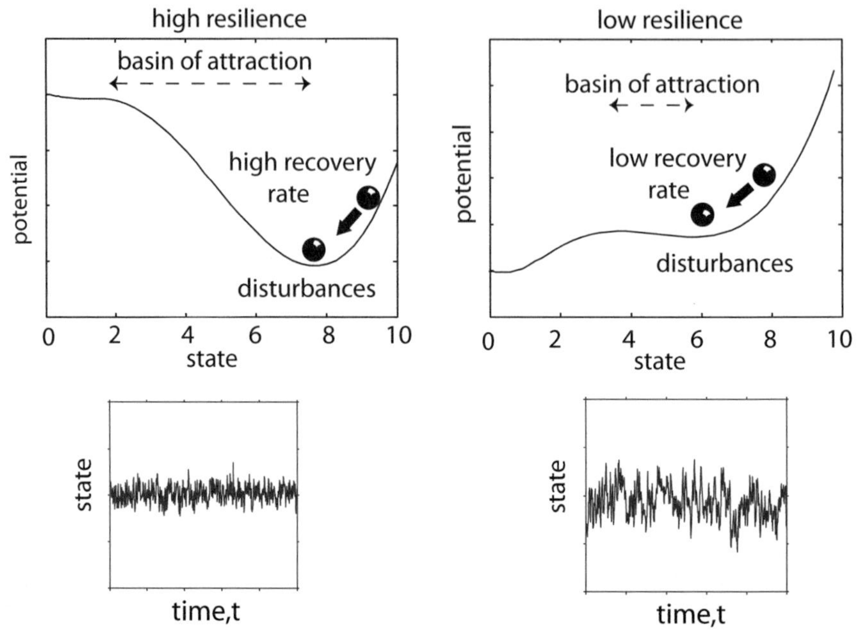

Fig. 9.12 The simple model which assimilates the tipping point to an attraction basin. The bottom inserts give the amplitudes of the disturbance

more complex kind of cell with a membrane-enclosed nucleus). Nick Bostrom has emphasized this point arguing that the absence of life on other planets is a sign that the Great Filter is beyond us because earthling have passed this critical step. However the Great Filter could be still in our future and take the form of global catastrophe like a nuclear war. These future events do not include natural disasters (like asteroid impact) because following Bostrom *even if they destroyed a significant number of civilizations we would expect some civilizations to get lucky and escape disaster, and some of these civilizations could then go on to colonize the universe*. So the question whether the Great Filter is ahead or beyond us is crucial and strangely enough could be resolved by looking at extrasolar planets in the realm of astrobiology. Jacob Haqq-Misra and coworkers made the consideration that the number of exoplanets N_{ec} that reveal evidence of feature x with confidence level c is given by

$$N_{ec} = \frac{\log(1-c)}{\log(1-x)} \tag{9.66}$$

For example x is the fraction of planets which develop life η_{life} and this amounts to 5% then the number of planets with confidence level of 95% is given by

$$N_{ec} = \frac{\log(0.05)}{\log(0.95)} \approx 60$$

Relation (9.66) comes from the fact that the probability P of x successes out of n tries, if each try has an independent probability p is given by the binomial distribution

$$P(x, n, p) = \frac{n!}{x!(n-x)!} P^x (1-p)^{n-x} \tag{9.67}$$

Now we suppose that $p = \epsilon \eta_x$ where p is the probability of detecting the feature, η_x is the statistical fraction of exoEarth candidates that exhibit the feature, and ϵ is our detection efficiency of the feature. Setting x = 0 and assuming $\epsilon \approx 1$, we find

$$P(0, N_{ec}, \eta_x) = (1 - \eta_x)^{N_{ec}} \tag{9.68}$$

And with $c = 1 - P$ we have (9.66). Now we define η_{tech} as the fraction of inhabited planets that show observable techno-spheres (for example, greenhouses gases), so the number of habitable planets we must observe in order to determine the prevalence of technology in the galaxy is

$$N_{tech} = \frac{\log(1-c)}{\eta_{life} \log(1 - \eta_{tech})} \tag{9.69}$$

If we assume the same 60 previous candidates, we have

$$\log(1 - \eta_{tech}) = \frac{(1-c)}{60 \times 0.05}, \quad \Rightarrow \eta_{tech} \approx 0.64$$

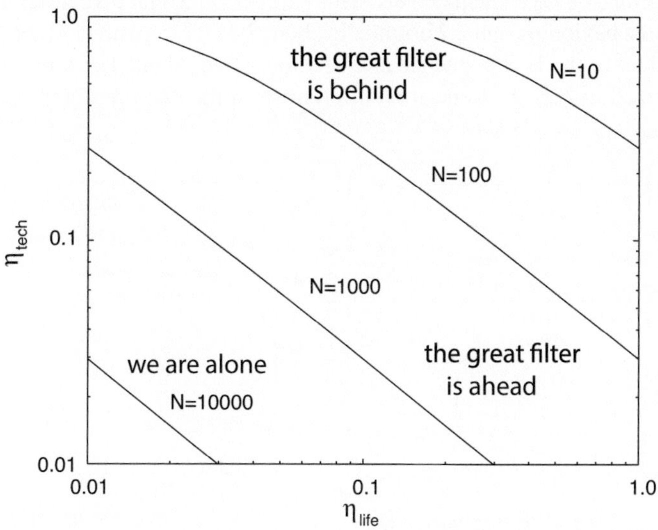

Fig. 9.13 The relation between the fractions η_{tech} and η_{life} once the total number of habitable planets N is fixed

We can now devise a thought experiment by fixing N_{tech} and the confidence level and then plot the resulting expression as in Fig. 9.13. This gives us the number of planet to explore to establish either η_{life} or η_{tech}. The experiment tells us that if we find η_{life} and η_{tech} very low that means we are alone (or almost) in the universe. If the results are that $\eta_{tech} < \eta_{life}$ means that the Great Filter could be in our future while the opposite would be in the case $\eta_{tech} > \eta_{life}$.

9.9 The Future of the Atmosphere and Biosphere

It is time to summarize what we would expect in our future. We have touched several times this argument from the seminal work of Caldeira and Kasting to a more stringent considerations we made about tipping points and great filter. Although there is some element which may predict the occurrence of the former substantially try remain unpredictable and the possibility to recover from a tipping point are important even with long time constant. Paradoxically the only predictable future is related to the evolution of the Sun because as a star it follows the Hertzsprung–Russell sequence. That means that in the first 12 billion years from its birth the luminosity of the Sun can be approximated by a relation like

$$L(t) = \frac{L_{\odot}}{[1 + 0.36(1 - t/t_{\odot})]} \tag{9.70}$$

where t is expressed in billion of year, and $L_{\odot}$ and $t_{\odot}$ are the present luminosity and age of the Sun (4.8 Gyr), respectively. This expression would give a star about 30% dimmer at the beginning and 2.2 brighter in about 7.2 Gyr from now. At that epoch the Sun would leave the H-R sequence and become a Red Giant. The luminosity curve is summarized in Fig. 9.14 and it is different from the one presented before when

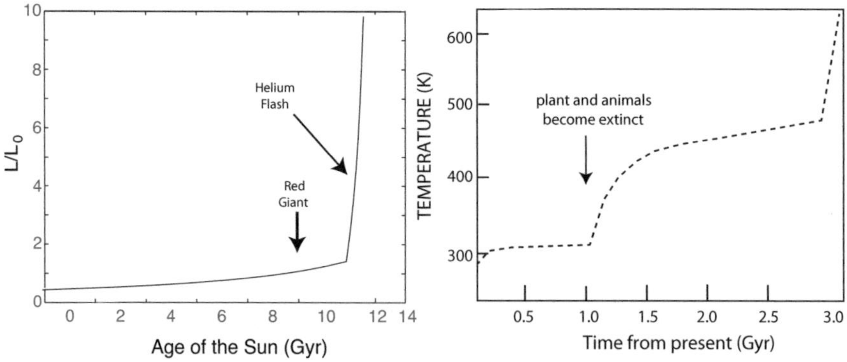

Fig. 9.14 On the left we show the Sun luminosity of function of time (the present epoch is about 4.8 Gyr). On the right the possible evolution of temperature

calculating the lifespan of the biosphere. Even more difference are present when the surface temperature calculation takes into account the moist greenhouse leading to the runaway greenhouse. The moist greenhouse may start around 330 K about 1Gyr from now while the runaway requires temperature around 420K and may start 2.8 Gyr from present time. The evaporation of the ocean reduces strongly the heat capacity of the surface of the Earth. Once the Sun becomes a Red Giant its diameter increases to the point of gobbling the planet Mercury and possibly even Venus. The Earth would be spared but the intense radiation would destroy any remaining life form.

Appendix

The Revelle Factor

Suppose we change the pressure of $p(CO_2)$ in the atmosphere by an amount $\Delta p(CO_2)$ then the total carbon dissolved DIC changes by ΔDIC. We define *Revelle factor* ϵ as the ratio of the relative changes of pCO_2 and DIC

$$\epsilon = \frac{\frac{\Delta p(CO_2)}{p(CO_2)}}{\frac{\Delta DIC}{DIC}} \tag{9.71}$$

With numerical experiments like those illustrated in the Appendix in Chap. 7 it is possible to obtain theoretically the value of the Revelle factor or better yet from historical data it results that the factor is comprised between 8 and 10. That means that if the carbon dioxide in the atmosphere change of 1% then the change in the ocean is only of 0.1%.

The Lifespan of the Biosphere MATLAB

```
%program lifespan1
clear all
close all
temp(1)=288.15;
time(1)=0.;
p0=320;
patm(1)=p0;
for i=1:1599
time(i)=(i-1)*10;
solar(i)=1368/(1-0.38*time(i)/4.55e4);
albedo=1.4981-0.0065979*temp(i)+8.567e-6*temp(i)*temp(i);
teff1=((1-albedo)*solar(i)/(4*5.669e-8));
```

```
teff2=sqrt(teff1);
teff=sqrt(teff2);
psi=log10(1.e-6*patm(i));
dtemp1=4.895e7/(temp(i)*temp(i))-3.9787e5/temp(i)-6.7084/(psi*psi);
dtemp=815.17+dtemp1+73.221/psi-30882/(temp(i)*psi);
temp(i+1)=teff+dtemp;
pco2(i)=p0*exp(-(temp(i+1)-288.15)/4.11);
if pco2(i)>0.003125
patm(i+1)=p0*pco2(i);
else
patm(i+1)=1;
end
tcels=1-(((temp(i+1)-273.15)-25)/25)^2;
if tcels<0 | patm(i+1)<10
productivity(i)=0;
else
productivity(i)=2*tcels*(p0-10)/(300+p0-10);
end
```

The Hybrid Planet

In a paper published in 2017, Adam Frank and his coworkers described the Anthropocene epoch as an hybridization of life in Earth in the sense that human activities not only are responsible for new evolutionary pressure but new selected directions for novel planetary ecosystem functions. We report a simple MATLAB script developed for their model

```
[t,x]=ode45('frank',[0 200],[0.01, 0.01]);
yyaxis left;
plot(t,x(:,1))
yyaxis right;
plot(t,x(:,2),'--')
function xdot=frank(t,x)
xdot=zeros(2,1);
teta=0;
gamma=0.01;
wk1=x(1)*(1+teta*x(2));
wk2=1-x(1);
if(wk1<wk2)
xdot(1)=wk1-x(1)*x(2)^2;
else
xdot(1)=wk2-x(1)*x(2)^2;
end
xdot(2)=gamma*x(1);
```

Critical Slowing Down

We consider the simple dynamical system described by the equation

$$\frac{dx}{dt} = \gamma(x-a)(x-b) \tag{9.72}$$

This system has two equilibrium points $x_1 = a$, $x_2 = b$ of which one is stable and the other is unstable. Assuming x_1 be the stable one we perturb the system in such a way that $x = x_1 + \epsilon$ we get

$$\frac{d}{dt}(x_1+\epsilon) = f(x_1+\epsilon) \tag{9.73}$$

where $f(x_1) = \gamma(x_1+\epsilon-a)(x_1+\epsilon-b)$ To the first order, we get

$$\frac{d}{dt}(x_1+\epsilon) = f(x_1+\epsilon) \approx f(x_1) + \left.\frac{\partial f}{\partial x}\right|_{x_1} \epsilon \tag{9.74}$$

That is

$$\frac{d\epsilon}{dt} = \left.\frac{\partial f}{\partial x}\right|_{x_1} \epsilon = \lambda\epsilon \tag{9.75}$$

With eigenvalues

$$\lambda_1 = \left.\frac{\partial f}{\partial x}\right|_a = -\gamma(b-a), \qquad \lambda_2 = \left.\frac{\partial f}{\partial x}\right|_b = \gamma(b-a) \tag{9.76}$$

If b>a we see that the first solution is stable (λ_1 is negative) and the larger the basin (b-a) the shorter is the recovery rate.

A very similar situation can be found for the zero-dimensional energy balance climate model that can be formulated as

$$\frac{dT}{dt} = \frac{S_0}{4}(1-\alpha_p) - (A+BT) \tag{9.77}$$

The solution to this equation can be written as

$$A + BT_0 = \frac{S_0}{4} H_0[x_s(T_0)] \tag{9.78}$$

where $H_0[x_s(T_0)]$ is a quote complicated function of the sine of ice latitude x_s

$$H_0[x_s(T_0)] = a_i + (a_f - a_i)\left[x_s + \frac{S_2(x_s - x_s^3)}{2}\right] \tag{9.79}$$

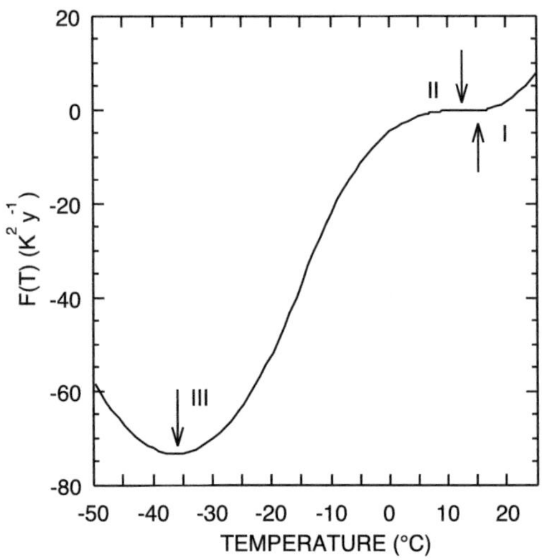

Fig. 9.15 The potential function of the zero-dimensional energy balance model. Notice the equilibrium points I, II, and III, of which II is unstable,

where a_i is the ice–albedo and a_f is the land–albedo while S_2 is related to Legendre polynomials. We now invent a potential function shown in Fig. 9.15.

$$F(T_0) = AT_0 + \frac{1}{2}BT_0^2 - \frac{S_0}{4}\int_0^{T_0} H(T_0')T_0' \tag{9.80}$$

whose minima are the solutions of (9.71)

$$\frac{dF}{dT_0} = 0, \quad \rightarrow A + BT_0 = \frac{S_0}{4}H_0[x_s(T_0)] \tag{9.81}$$

We can now try to find the response time of the two minima by considering a small deviation from the equilibrium

$$T(t) = T_0 + T' \tag{9.82}$$

Then differentiating (9.78) we have

$$\frac{dT'}{dt} = -BT' + q\left(\frac{\partial H_o}{\partial T_0}\right)T' \tag{9.83}$$

where $q = S_0/4$. Then we get

$$qH_0 = A + BT_0, \quad \Rightarrow H_0\frac{\partial q}{\partial T_0} + q\frac{\partial H_0}{\partial T_0} = B \tag{9.84}$$

which substituted in (9.83) gives

$$\frac{\partial T'}{\partial t} = -H_0 \frac{\partial q}{\partial T_0} T' \tag{9.85}$$

which tells you that for stability $\partial q / \partial T_0 > 0$. This corresponds to a time constant

$$\frac{1}{\tau} \approx H_0 \frac{\partial q}{\partial T_0} \tag{9.86}$$

Now for a change of $\Delta S_0 \approx 0.02 \times 340 \approx 7wm^{-2}$ we have about 3.5 K change in temperature so that

$$\frac{\Delta q}{\Delta T_0} \approx 2wm^{-2}K^{-1}.$$

And assuming that for the warm solution $H_0 \approx 0.3$ and for the cold 0.6 we have that for the cold solution a much shorter response time.

References

Abe, Y., et al.: Habitable zone limits for dry planets. Astrobiology **11**, 443 (2011)

Allen, M.R., Frame, D.J.: Warming caused by cumulative carbon emissions towards the trillionth tonne. Nature **458**, 1163 (2009)

Archer, D., et al.: Multiple timescale for neutralization of fossil fuel CO_2. Geophys. Res. Lett. **24**, 405 (1997)

Armstrong McKay, D.I., Staal1, A., et al.: Exceeding 1.5°C global warming could triggern multiple climate tipping points. Science **377**, 1171 (2022)

Ashwin, P., von der Heydt, A.S.: Extreme sensitivity and climate tipping points. J. Stat. Phys. **179**, 1531 (2020)

Bostrom, N.: Where are they? MIT Technol. Rev. (2008)

Bounama, C., et al.: The fate of Earth's ocean. Hydrol. Earth Syst. Sci. **5**, 569 (2001)

Caesar, L., McCarthy, G.D., et al.: Current Atlantic meridional overturning circulation weakest in last millennium. Nature Geos. **14**, 118 (2021)

Caldeira, K., Kasting, J.F.: The Life Span of the Biosphere Revisited, vol. 360, p. 721 (1982)

Dakos, V., Scheffer, M., et al.: Slowing down as an early warning signal for abrupt climate change. PNAS **105**, 14308 (2008)

Davies, G.F.: Thermal histories of convective earth models and constraints on radiogenic heat production in the earth. J. Geophys. Res. **85**, B5, 2517 (1980)

Franck, S., et al.: Reduction of biosphere life span as a consequence of geodynamics. Tellus **52B**, 94 (2000)

Franck, S., Bounama, C., von Bloh, W.: Causes and timing of future biosphere extinctions. Biogeosciences **3**, 85 (2006)

Frank, A., et al.: Earth as a hybrid planet: the anthropocene in an evolutionary astrobiological context. Anthropocene **19**, 13 (2017)

Goldblatt, C., Robinson, T.D., et al.: Low simulated radiation limit for runaway greenhouse climates. Natgeo **6**, 611 (2013)

Graham, R.J., Halevy, I., et al.: Substantial extension of the lifetime of the terrestrial biosphere. Planet. Sci. J. **5**, 255 (2024)

Hanson, R.: The great filter—are we almost past it? Arch. Orig. https://mason.gmu.edu/~rhanson/greatfilter.html

Karato1, S.: Deep mantle melting, global water circulation and its implications for the stability of the ocean mass. Prog. Earth Planet. Sci. **7**, 76 (2020)

Krissansen-Totton, J., Catling, D.C.: Constraining climate sensitivity and continental versus seafloor weathering using an inverse geological carbon cycle model. NatCo **8**, 15423 (2017)

Lenton, T.M., von Bloh, W.: Biotic feedback extends the life span of the biosphere. Geophys. Res. **28**, 1715 (2001)

Lovelock, J.E., Whitfield, M.: Life span of the biosphere. Nature **296**, 561–563 (1982)

Matthews, H.D., Caldeira, K.: Stabilizing climate requires near-zero emissions. Geophys. Res. Lett. **35**, L04705 (2008)

Montenegro, A., Brovkin, V., et al.: Long term fate of anthropogenic carbon. Geophys. Res. Lett. **34**, L19707 (2007)

O'Malley-James, J.T.: Swansong Biospheres: refuges for life and novel microbial biospheres on terrestrial planets near the end of their habitable lifetimes. Int. J. Astrobiol. **12**, 99 (2013)

O'Malley-James, J.T.: Swansong Biospheres II: the final signs of life on terrestrial planets near the end of their habitable lifetimes. Int. J. Astrobiol. **13**, 229 (2014)

Ozaki, K.C.T.: Reinhard The future lifespan of Earth's oxygenated atmosphere. Nat. Geosci. **14**, 138 (2021)

Ramanathan, V., Feng, Y.: On avoiding dangerous anthropogenic interference with the climate system: formidable challenges ahead. PNAS **105**, 14425 (2008)

Rybicki, K.R., Denis, C.: On the final destiny of the earth and the solar system. Icarus **151**, 130 (2001)

Savitch, E., et al.: Triggering a climate change dominated "Anthropocene": is it common among exocivilizations? Astron. J. **162**, 196 (2021)

Scheffer, M., Bascompte, J., et al.: Early-warning signals for critical transitions. Nature **461**, 53 (2009)

Solomon, S., et al.: Irreversible climate change due to carbon dioxide emissions. PNAS **106**, 1704 (2009)

Steffen, W., Rockström, et ak. J.: Trajectories of the Earth System in the Anthropocene. PNAS **115**, 8252 (2019)

Stocker, T.: Introduction to Climate Modelling. Springer (2011)

Stocker, T.F.: The closing door of climate targets. Science **339**, 280 (2013)

Zaliapin, I., Ghil, M.: Another look at climate sensitivity. Nonlinear Process. Geophys. **17**, 113 (2010)

Index

G. Visconti, *Origin and Evolution of the Atmosphere and Ocean*,
https://doi.org/10.1007/978-3-032-02653-8

MIX
Papier aus verantwortungsvollen Quellen
Paper from responsible sources
FSC® C105338

If you have any concerns about our products,
you can contact us on
ProductSafety@springernature.com

In case Publisher is established outside the EU,
the EU authorized representative is:
Springer Nature Customer Service Center GmbH
Europaplatz 3, 69115 Heidelberg, Germany

Printed by Libri Plureos GmbH
in Hamburg, Germany